T0406752

Springer Water

More information about this series at http://www.springer.com/series/13419

Sangam Shrestha

Climate Change Impacts and Adaptation in Water Resources and Water Use Sectors

Case studies from Southeast Asia

Sangam Shrestha
Water Engineering and Management
Asian Institute of Technology
Klong Luang
Thailand

ISBN 978-3-319-09745-9 ISBN 978-3-319-09746-6 (eBook)
DOI 10.1007/978-3-319-09746-6

Library of Congress Control Number: 2014945772

Springer Cham Heidelberg New York Dordrecht London

Printed on acid-free paper

Springer is part of Springer Science+Business Media (www.springer.com)

Foreword

The threat to sustainability of water resources—one of the most vital of all natural resources—is becoming more and more severe by the day, caused by a variety of anthropogenic activities and natural phenomena. There is a growing body of literature which suggests that this threat will further intensify under the effects of climate change. Recent research reveals that even if all emissions were stopped now, future climate will still be warmer than the pre-Industrial Revolution levels because the greenhouse gases already emitted are likely to persist in the atmosphere for thousands of years. Hence, understanding the potential impacts of climate change on water resources and water use sectors, and developing appropriate adaptation options is the need of the hour.

This book, *Climate Change Impacts and Adaptation in Water Resources and Water Use Sectors*, provides in-depth and comprehensive knowledge about various techniques to analyze the impact of climate change on water resources using contemporary climate and hydrological models. Further, through the use of case (research) studies, a step-by-step procedure has been illustrated to evaluate the impacts of climate change on water resources and selected water use sectors like agriculture. The level of details provided for each case study will provide readers with enough insight to replicate this work in diverse settings. Additionally, useful information can be obtained on developing adaptation options in selected water use sectors in order to counter the effects of climate change.

The author, Sangam Shrestha, has considerable research experience in climate change impact assessment and adaptation in the water sector in South and Southeast Asia, particularly on hydrology; crop production; and evaluating adaptation measure to offset the negative impacts. Apart from holding a faculty position at the Asian Institute of Technology (AIT), he is also a Research Fellow at the Institute for Global Environmental Strategies (IGES), Japan. Climate change adaptation in the water sector is a key area of scientific and policy research at IGES, and over the years significant strides have been made by the institute in informing judicious decision support tools to foster the sustainability of water resources in the region and beyond. At IGES we hope to work as an "agent of change," facilitating the transition to a sustainable society and improving the well-being of people in the region.

This book is an excellent resource for students, researchers, and water managers. Further, the implications of the outcomes of the various case studies will be of particular interest to decision and policy makers. The importance, and need, of an effective water management system in the existing times cannot be overstated, particularly in Southeast Asia which, among the regions in the world, is one of the most vulnerable to the impacts of climate change. This book can be a useful tool for addressing this imminent need.

Hideyuki Mori
President
Institute for Global Environmental Strategies

Acknowledgments

A number of individuals have contributed to the preparation of this book. I extend my deepest gratitude to all of them, but because of space constraints it is not possible to mention all the names here. However, it would be injustice if I failed to mention a few individuals whose contributions are particularly significant. My sincere thanks to my students: Birat Gyawali, Bui Thi Thu Trang, Chatchai Chingchanagool, Naw May Mya Thin, and Worapong Lohpaisankrit for their effort to conduct research under my supervision.

Special thanks are also due to Proloy Deb who helped in collecting the data and information needed to prepare book. I would also like to thank Smriti Malla for her tireless efforts in compiling and formatting all the chapters.

Contents

About the Author

Dr. Sangam Shrestha is an Assistant Professor of Water Engineering and Management at Asian Institute of Technology (AIT), Thailand. He is also a Visiting Faculty of University of Yamanashi, Japan and Research Fellow of Institute for Global Environmental Strategies (IGES), Japan. His research interests are within the field of hydrology and water resources including the climate change impact assessment and adaptation on water sector, water footprint assessment, and groundwater assessment and management.

After completing his Ph.D., Dr. Shrestha continued his postdoctoral research in the GCOE project of University of Yamanashi in Japan until 2007 where he was involved in development and application of material circulation model and groundwater research in the Kathmandu Valley. He then worked as a policy researcher at Institute for Global Environmental Strategies (IGES) where he was actively involved in research and outreach activities related to water and climate change adaptation and groundwater management in Asian cities. Dr. Shrestha has published more than two dozen peer-reviewed international journal articles and presented more than three dozen conference papers ranging from hydrological modeling to climate change adaptation in the water sector. His recent publication includes "Kathmandu Valley Groundwater Outlook" and "Climate Change and Water Resources."

His present work responsibilities in AIT include delivering lectures at the postgraduate and undergraduate levels, supervising research to postgraduate students, and providing consulting services on water-related issues to government and donor agencies and research institutions. He has been conducting several projects related to water resources management, climate change impacts, and adaptation being awarded from International organizations such as APN, CIDA, EU, FAO, IFS, IGES, UNEP, and UNESCO.

Abbreviations

ADB	Asian Development Bank
APN	Asia-Pacific Network for Global Change
AR5	Fifth Assessment Report
Cal	Calibration
CDC	Canopy Decline Coefficient
CGC	Canopy Growth Coefficient
CICS	Canadian Institute for Climate Studies
CV	Coefficient of Variation
DAT	Days After Transplanting
DEM	Digital Elevation Model
DHI	Danish Hydraulic Institute
ECHAM5	European Centre-Hamburg Model Version 5
EI	Efficiency Index
EV	Extreme Value
FAO	Food and Agriculture Organization
FC	Field Capacity
GCM	General Circulation Model
GDP	Gross Domestic Product
GHGs	Greenhouse Gases
HadCM3	Hadley Centre Coupled Model Version 3
HEC-HMS	Hydrologic Engineering Center's Hydrologic Modeling System
HRUs	Hydrological Response Units
HYV	High Yielding Variety
IFPRI	International Food Policy Research Institute
IGBP	International Geosphere Biosphere Programme
IPCC	Intergovernmental Panel on Climate Change
IWR	Irrigation Water Requirement
MCM	Million Cubic Meters
MPI-OM	Max Planck Institute Ocean Model
NAM	Nedbør-Afstrømnings-Model
NCEP	National Centers for Environmental Prediction

NIP	Ngamoeyeik Irrigation Project
PRECIS	Providing Regional Climates for Impact Studies
RCDB	Randomized Complete Block Design
RCM	Regional Climate Model
RID	Royal Irrigation Department
RMSE	Root Mean Square Error
SD	Standard Deviation
SDSM	Statistical DownScaling Model
SEA	Southeast Asian
SEA START	Southeast Asian System for Analysis, Research and Training
SLE	Sea Level Equivalent
SRES	Special Report on Emission Scenarios
SRTM	Shuttle Radar Topography Mission
TAW	Total Available Water
Tmax	Maximum Temperature
TMD	Thai Meteorological Department
Tmin	Minimum Temperature
UHM	Unit Hydrograph Method
UNDP	United Nations Development Program
Val	Validation
VB	Volume Bias

Chapter 1
Introduction

Climate change has been defined as a statistically significant variation of either variability or the mean of climate enduring for a prolonged period, generally decades or longer (Cubasch et al. 2013). Although there are both natural and anthropogenic causes of climate change, yet the contribution of human factors has been identified as more remarkable (Power and Goyal 2003). Industrial revolution, agricultural expansion, burning of fossil fuel and transportation are the major drivers eventually lead to injection of chlorofluorocarbons (CFCs) and other green house gases (GHGs) in the atmosphere. The GHGs act as accelerator which traps the incoming heat from solar radiation. 81 % of total radiative forces in the atmosphere are contributed by CO_2 and CH_4 (Tuckett 2009). The increase in the radiative forces in atmosphere has lead to increased temperature (IPCC 2007; Ding et al. 2006) and has significantly altered precipitation patterns (Gao et al. 2012; Liang et al. 2011). Altercation of these two meteorological variables in hydrological cycle has significantly affected other components directly and indirectly (Fig. 1.1) conclusively affecting agricultural and water sectors (Vargas-Amelin and Pindado 2013; Babel et al. 2011). The observed climate change impacts at global scale are detrimental and it has affected several sectors. The major ones include global average air and ocean temperatures, spatial extent of snow and ice, glacial retreat and increased sea level. Other detected effects include higher intensities of extreme events such as floods and droughts, intense hotter days and cyclonic activities which ultimately have significant influence on agricultural production and human lives.

The Intergovernmental Panel on Climate Change (IPCC)'s Fifth Assessment Report (AR5) suggests the current climate is changing at a significant rate relative to earlier periods. A superficial relationship derived from the temperature rise and sea level rise for 120 years reflects a potential increase of sea level by more than one meter by 2,100 (Rayanakorn 2011). Other observed effect includes the changes in cryosphere which has been documented virtually in all cryospheric components with robust evidence that are in general a response to the reduction of snow and ice masses due to enhanced warming. Although the changes in mountain glaciers and

S. Shrestha, *Climate Change Impacts and Adaptation in Water Resources and Water Use Sectors*, Springer Water, DOI 10.1007/978-3-319-09746-6_1

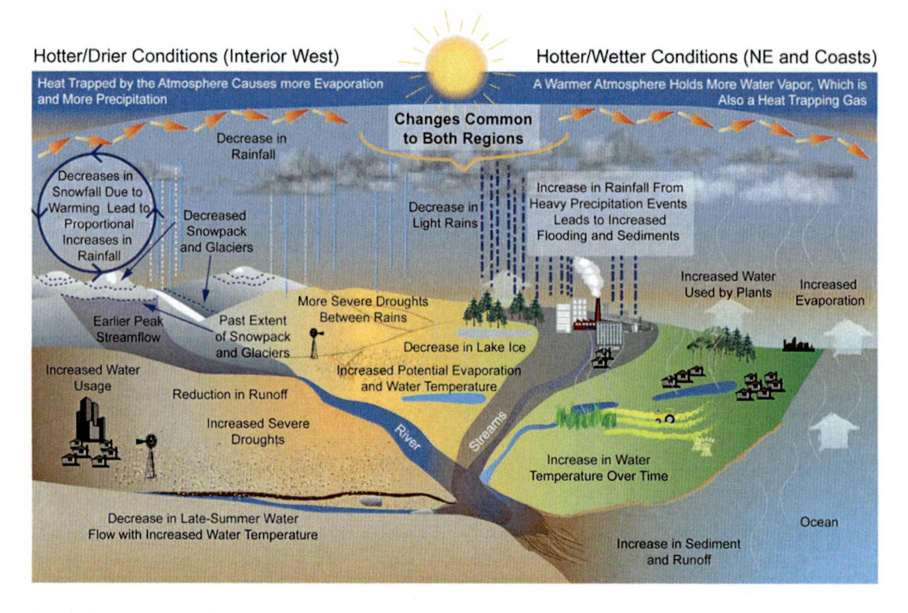

Fig. 1.1 Interdependence of meteorological variables and hydrological cycle. *Source* USGCRP 2009

ice caps have been well documented in runoff estimation (Box et al. 2006), altering the hazard magnitude and intensity (Haeberli and Burn 2002) and ocean freshening (Bindoff et al. 2007), there is also a contemporary evidence of the crustal uplift in response to recent glacier melting in northern America. In general, it is observed that since 1960s, loss in glaciers outside Greenland and Antarctica are 0.76 mm yr^{-1} sea level equivalent (SLE) during 1993–2009 and a significant increase of 0.83 mm yr^{-1} for the period 2005–2009 (IPCC 2013). Similarly the observed changes for sea ice extent illustrate a significant declination of 3.8 % per decade with larger losses in summer and autumn relative to other seasons. Recent studies also validated climate change driven by various GHGs has significantly affected the hydrologic regime. Shorter wet season with intense rainfall and extended dry season not only has affected the water availability, water distribution but also the agricultural planning and management (Seiller and Anctill 2013; Kang and Khan 2009). Especially in case of the delta, where the ecosystem is highly threatened by not only the direct climatic factors, but also the indirect implications of the sea level rise, backward flow of sea water to river and salinity intrusion.

It is anticipated that climate change will impose many risks and vulnerability for the present and the future generation. The most susceptible being the developing or the least developed nations which posses very less coping capacity to the negative effects of climate change (Rayanakorn 2011). The key findings of the latest AR5 report suggest that under climate change, several ecosystems are expected to be threatened with possible extinction of many plant and animal species. Historical observations of agricultural productivity at global scale reflect higher productivity in higher latitudes and a relative lower yield for regions near to equator and this difference is expected to get widen in the future under climate change (Ray et al. 2012). The current increasing population and human dwelling in the flood plains and coastal areas have increased the risk of the people vulnerable to climatic events. Furthermore, spread of various infectious diseases also has been attributed to climate change with a confidence level ranging from moderate to high (St Laurent and Mazumder 2014; El-Fadel et al. 2012).

In order to appraise the implications of climate change on these sectors, globally impact assessment studies has been carried out by the projections done by General Circulation Models (GCMs). GCMs are crucial tools to reproduce the virtual future climate based on energy and mass balance equations (Chiew et al. 2013). They represent the physical processes in atmosphere, ocean, cryosphere and land surface based on response of global climate system to the increasing greenhouse gas concentrations depending of five selection criteria: Consistency with global projections, physical plausibility, applicability in impact assessments, representativeness and accessibility (IPCC 2007). GCMs illustrates the climate by using a three dimensional grid over globe (Fig. 1.2) with a horizontal resolution ranging from

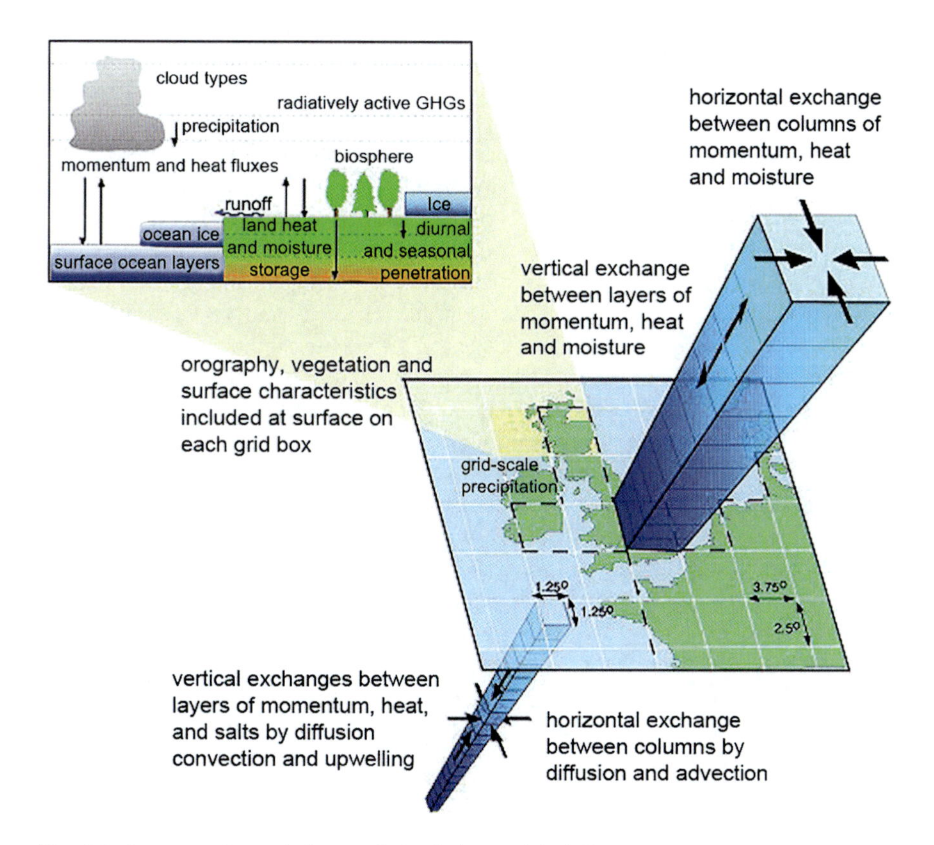

Fig. 1.2 Structure of a typical general circulation model (GCM)

250 to 600 km and 10 to 20 vertical layers in atmosphere and additional up to 10 layers in case of ocean. Due to the coarse spatial resolution relative to the exposure units, are not suitable for basin scale impact assessment studies (Zorita and von Storch 1999). In addition, many physical processes attributed to clouds also occur in smaller units and leads to uncertainty in modelling. Therefore for impact assessment studies, the outputs are necessarily converted or downscaled to specific target weather stations using the historic weather data or available database. The spatially refined multiple ensemble outputs of GCMs are further used in models for impact assessment studies along with their uncertainty.

The Southeast Asian (SEA) region comprises of 11 nations namely Myanmar, Thailand, Laos, Cambodia and Vietnam in mainland whereas Indonesia, Philippines, Brunei, East Timor, Malaysia and Singapore in the maritime region. The region serves as a homeland for approximately 618 million people and a land area of 4.5 million km^2. The Mekong River originating from China serves as the lifeline of the people in the mainland, providing livelihood to more than 70 million people along with the river basin. The current altercations in the climate pattern driving the water budget have affected various sectors in the region including energy distribution, agriculture and water supply. Moreover, the tropical and monsoonal countries in the lower Mekong basin also have encountered many frequent and intense floods in recent past namely Myanmar and Vietnam being the most vulnerable countries in the region. On contrary intense and longer spell of droughts have prevailed in Thailand, Laos and Cambodia in recent past (Hundertmark 2008) indicating the implications of climate change are ambiguous and alters with locations. This implies in order to appraise the impacts of climate change on various sectors for any region, studies must be focused at basin/provincial level however, in this book we have constrained ourselves on the mainland regions of the SEA region.

The main purpose of this book is to inform water managers and decision makers about the climate change, its impact on water resources and selected water use sectors and how to adapt to these changes. All the case studies presented in this book are based on established theories, principle and application of sound methodologies. The book starts with the Chap. 1 which provides the general background of climate change and its observer impacts on water resources and water use sectors. It also highlights the need of conducting climate change impact assessment at basin level to formulate adaptation strategies to cope with negative impacts of climate change on hydrology, irrigation water requirements and crop production.

The Chap. 2 focuses on the assessment of climate change impacts on water availability in Thailand. The result of the study shows that the water availability in the future decades varies for the dry and wet season. In case of dry season, the coastal areas show a decline in water availability in the near future then tending to increase to the similar amount as of current situation in the late part of century. However, in case of wet season an increasing trend of water availability is observed in future. Nonetheless, considering the whole country for dry season the water availability is expected to be decreased in the early part of the century followed by an increasing trend by the end of the century relative to present water availability for both scenarios. Similarly a univocal increasing trend of water availability is expected for wet season indicating the possibility increased frequency and intensity of floods.

The Chap. 3 aims to investigate the impact of climate change on the inflow to Ubolratana dam in Thailand. The outputs of regional climate model PRECIS was used to project the future climate under two emission scenarios A2 and B2. The result for A2 and B2 emission scenarios is divided into three periods of 2010–2039, 2040–2069 and 2070–2099. In future, the higher precipitation change was observed in the downstream area of the Chi river basin. The mean maximum temperature change of period 1–3 in range of +0.5 to +2.5 °C for A2 and +0.5 to +1.9 °C for B2 scenarios. Similarly, the mean minimum temperature change in range of +0.6 to +3. 1°C for A2 and +0.6 to 2.3°C for B2 scenarios. MIKE11 NAM model was used to compute the inflow to Ubolratana dam of Chi river basin. Simulation results showed that inflows will increase in future for both A2 and B2 scenarios. It was also found that with the increasing inflows, the resiliency and reliability of the dam will be decreased whereas its vulnerability will be increased.

The Chap. 4 analyze the impacts of climate change on flood hazard in Yang River Basin under future climatic scenarios with coupling of a physically-based distributed hydrological model, Block-wise application of TOPMODEL using Muskingum-Cunge flow routing (BTOPMC) and hydraulic model, HEC-RAS. Results indicate that croplands are being mostly affected by 100 year return period in case of baseline period. The probable increase in flood hazard under climate change scenarios threatens the increased inundation of croplands area and indicates the potential damage in food production and its impacts on livelihood of local people.

The Chap. 5 analyzes the temporal impacts of climate change on irrigation water requirement (IWR) and yield for rainfed rice and irrigated paddy respectively at Ngamoeyeik Irrigation Project (NIP) in Myanmar. Climate projections from two General Circulation Models (GCMs) namely ECHAM5 (scenario A2 and A1B) and HadCM3 (scenarios A2 and B2) were derived for NIP for future time windows (2020s, 2050s and 2080s). The analysis shows a decreasing trend in maximum temperature (−0.8 to +0.1°C) for the three scenarios and three time windows considered; however, an increasing trend is observed for minimum temperature (+0. 2–+0.4°C) for all cases. The analysis on precipitation also suggests that rainfall in wet season is expected to vary largely from −29 % (2080s; A1B) to +21.9 % (2080s; B2) relative to the average rainfall of the baseline period. A decreasing trend of irrigation water requirement is observed for irrigated paddy in the study area under the three scenarios indicating that small irrigation schemes are suitable to meet the requirements. An increasing trend in the yield of rainfed paddy was estimated under climate change demonstrating the increased food security in the region.

In Chap. 6, several adaptation measures were evaluated to overcome the negative impact of climate change on rice production in the Quang Nam province of Vietnam. Results show that climate change will reduce rice yield from 1.29 to 23.05 % during the winter season for both scenarios and all time periods, whereas an increase in yield by 2.07–6.66 % is expected in the summer season for the 2020s and 2050s; relative to baseline yield. The overall decrease of rice yield in the winter season can be offset, and rice yield in the summer season can be enhanced to potential levels by altering the transplanting dates and by introducing supplementary irrigation. Late transplanting of rice shows an increase of yield by 20–27 % in future. Whereas supplementary irrigation of rice in the winter season shows an increase in yield of up to 42 % in future. Increasing the fertilizer application rate enhances the yield from 0.3 to 29.8 % under future climates. Similarly, changing the number of doses of fertilizer application increased rice yield by 1.8–5.1 %, relative to the current practice of single dose application. Shifting to other heat tolerant varieties also increased the rice production. Based on the findings, changing planting dates, supplementary irrigation, proper nutrient management and adopting to new rice cultivars can be beneficial for the adaptation of rice cultivation under climate change scenarios in central Vietnam.

The overall conclusion of these six case studies is that climate change will have impacts on water resources and water use sectors such as agriculture. However the magnitude and nature of the impacts varies with location and time period. Therefore it is very essential to formulate the adaptation strategies at local level to offset the negative impacts or to exploit the beneficial opportunities from climate change.

References

Babel MS, Agarwal A, Swain DK, Herath S (2011) Evaluation of climate change impacts and adaptation measures for rice cultivation in Northeast Thailand. Clim Res 46:137–146

Bindoff NL, Willebrand J, Artale V, Cazenave A, Gregory J, Gulev S, Hanawa K, Le Quéré C, Levitus S, Nojiri Y, Shum CK, Talley LD, Unnikrishnan A (2007) Observations: oceanic climate change and sea level. In: Solomon S, Qin D, Manning M, Chen Z, Marquis M, Averyt KB, Tignor M, Miller HL (eds) Climate change 2007: the physical science basis. Contribution of working group I to the fourth assessment report of the intergovernmental panel on climate change. Cambridge University Press, Cambridge

Box JE, Bromwich DH, Veenhuis BA, Bai LS, Stroeve JC, Rogers JC, Steffen K, Haran T, Wang SH (2006) Greenland ice sheet surface mass balance variability (1998–2004) from calibrated polar MM5 output. J Clim 19(2):2783–2800

Chiew FHS, Kirono DGC, Kent DM, Frost AJ, Charles SP, Timbal B, Nguyen KC, Fu G (2013) Comparison of runoff models using rainfall from different downscaling methods for historical and future climates. J Hydrol 387:10–23

Cubasch U, Wuebbles D, Chen D, Facchini MC, Frame D, Mahowald N, Winther JG (2013) Introduction. In: Stocker TF, Qin D, Plattner GK, TIgnor M, Allen SK, Boschung J, Nauels A, Xia Y, Bex V, Midgley PM (eds) Climate change 2013: the physical science basis. Contribution of working group I to the fifth assessment report of the intergovernmental panel on climate change. Cambridge

Ding YH, Ren GY, Shi GY, Gong P, Zheng XH, Zhai PM, Zhang DE, Zhao ZC, Wang SW, Wang HJ, Luo Y, Chen DL, Gao Xj, Dai XS (2006) National assessment report of climate change (I): climate change in China and its future trend. Advan Climate Change Res 2(1):3–8

El-Fadel M, Ghanimeh S, Maroun R, Alameddine I (2012) Climate change and temperature rise: implications on food- and water-borne diseases. Sci Total Environ 437:15–21

Gao F, Yu Z, Duan J, Ju Q (2012) Impact of climate change on water resources at local area in Anhui province. Procedia Engineering 28:319–325

Haeberli W, Burn C (2002) Natural hazards in forests: glacier and permafrost effects as related to climate change. In: Sidle RC (ed) Environmental change and geomorphic hazards in forests. Wallingford, Oxon, CABI Publishing, pp 167–202

Hundertmark W (2008) Building drought management capacity in the Mekong River Basin. Irri Drain 57:279–287

IPCC (2007) The physical science basis. In: Solomon S, Qin D, Manning M, Marquis M, Averyet K, Tignor MM et al. (eds) Contribution of working group I to the fourth assessment report of the intergovernmental panel on climate change. Cambridge University Press, Cambridge

Kang Y, Khan S (2009) Climate change impacts on crop yield, crop water productivity and food security: a review. Prog Nat Sci 19(12):1665–1674

Liang LQ, Li LJ, Liu Q (2011) Precipitation variability in Northeast China from 1961 to 2008. J Hydrol 404:67–76

Power HC, Goyal A (2003) Comparison of aerosol and climate variability over Germany and South Africa. Int J Climatol 23:921–941

Ray DK, Ramankutty N, Mueller ND, West PC, Foley A (2012) Recent patterns of crop yield growth and stagnation. Nat Commun. doi:10.1038/ncomms2296

Rayanakorn K (2011) Climate change challenges in the Mekong region. Chiang Mai University Press, Chiang mai, Thailand

Seiller G, Anctill F (2013) Climate change impacts on the hydrologic regime of a Canadian river: comparing uncertainties arising from climate natural variability and lumped hydrological model structures. Hydrol Earth Syst Discuss 10:14189–14227

St Laurent J, Mazumder A (2014) Influence of seasonal and inter-annual hydro-meteorological variability on surface water fecal coliform concentration under varying land-use composition. Water Res 48:170–178

Tuckett RP (2009) The role of atmospheric gases in global warming. In: Climate change observed impacts on planet earth. Elsevier. pp 3–19

Vargas-Amelin, Pindado P (2013) The challenge of climate change in Spain: water resources, agriculture and land. J Hydrol. In Press

Zorita E, von Storch H (1999) The analogue method as a simple statistical downscaling technique: comparison with more complicated methods. J Clim 12:2474–2489

Chapter 2
Assessment of Water Availability Under Climate Change Scenarios in Thailand

Abstract This paper investigates the potential impact on climate change on future water availability in Thailand. For this study, entire country was divided into nine Hydrological Response Units (HRUs) and the hydrological modeling was performed by Hydrologic Engineering Center's Hydrologic Modeling System (HEC-HMS) for each HRU using the future decadal climate data obtained from the Regional Climate Model (RCM) named Providing Regional Climates for Impact Studies (PRECIS) which was further bias corrected by using ratio method for two emission scenarios A2 and B2. The simulation shows that the water availability in the future decades at the different HRUs varies for the dry and wet season. In case of dry season, the coastal HRUs show a decline in water availability in the near future then tending to increase to the similar amount as of current situation in the late part of century. However, in case of wet season all the HRUs shows increasing trend of water availability in future. Nonetheless, considering the whole country for dry season the water availability is expected to be decreased in the early part of the century followed by an increasing trend by the end of the century relative to present water availability for both scenarios. Similarly a univocal increasing trend of water availability is expected for wet season indicating the possibility increased frequency and intensity of floods.

Keywords Climate change · HEC-HMS · PRECIS · Water availability · Thailand

2.1 Introduction

Southeast Asia is expected to be seriously affected by the impacts of climate change due to the high dependency of economy on agriculture and water resources in the region (IPCC 2007). The region's water resource is already affected by the rapid population growth, urbanization, agricultural and hydropower demand. Recent extreme events in Thailand shows it is under water crisis, in addition the intensity of the extreme events are also expected to increase in the future (Graiprab et al. 2010). Two most important problems attributed by climate change in the region are floods

S. Shrestha, *Climate Change Impacts and Adaptation in Water Resources and Water Use Sectors*, Springer Water, DOI 10.1007/978-3-319-09746-6_2

and droughts (SEA START RC 2009). Flooding negatively affects the crops, livelihoods and infrastructure throughout the country whereas drought affects the crop production specifically in the Northeast region (Kranz et al. 2010). Similarly, studies show that the impact of climate change are regional and its affects are also concentrated at regional scale (Chiew et al. 2009; Dore 2005) although the water management policies target at national scale.

Climate change is anticipated to have significant alteration of the global water cycle through changes in temperature and precipitation (Sharma and Babel 2013). The change in precipitation regime, in terms of intensity and frequency inclusive of spatial distribution has already been reported worldwide (Jiang et al. 2007a; Dore 2005). Contemporary studies of state of art on climate change impacts on hydrology in various watersheds in the world validate changes in the annual and seasonal pattern of flows (Li et al. 2013; Boyer et al. 2010; Jiang et al. 2007b). The increasing demand of freshwater by virtue of factors such as population growth and land use change has forced water resources under threat. Additionally, climate change has rendered its availability in the future towards more uncertainty (Davis and Simonovic 2011).

The Third Assessment Report of the Intergovernmental Panel on Climate Change (IPCC) released in 2001 reported intensification of the global hydrological cycle with its implications on surface and groundwater resources (IPCC 2001). Although several studies (Babel et al. 2013; Nohara et al. 2006; Arnell 2003) have been conducted to understand the impacts on river runoff by using the advanced global hydrological models that are driven by ensembles of climate models yet the influence of climate change at national scale and its variation with basin scale is still under dilemma (Minville et al. 2008).

For the last two decades GCMs have confirmed to be an essential tool for climate change impact assessment studies (Weart 2010). Although the simulated scenarios are advisable for the regional to national scale studies, they are less suitable for basin level studies due to their coarse spatial resolution. Several techniques have been developed to overcome this issue but still there is a demand to further develop the existing methods for impact assessment studies. Bias correction has been successfully applied in many parts of world for linking GCMs and hydrological models of impact assessment (Koutroulis et al. 2013; Leander and Buishand 2007). In addition although several hydrological models are available, HEC-HMS is a versatile semi-distributed model and its performance has been accepted in many basins in the world (Chu and Steinman 2009; Sharma et al. 2007).

Despite of the significant progress on the basin level climate change impacts assessment studies, a comprehensive study comprising of basin scale study attributing to national level water availability is necessary for Thailand. With limited adaptive capacity, the people are expected to be severely threatened by the additional influence of climate change. In order to address this issue, this paper presents the analysis of the future changes in local climate at hydrological response units (HRUs) and assesses their impact on national scale water availability to help in managing water resources more efficiently and prepare necessary plans for adaptation in changing climatic conditions.

2.2 Study Area

Thailand lies within 5°37′–20°28′ north latitude and 97°21′–105°38′ east longitudes. The study includes the 25 major river basins in Thailand covering an area of approx. 5,13,000 km^2. All major river basins were grouped into nine HRU (Hydrological Response Unit) based on the physiographic characteristics for easiness in hydrological modeling. Figure 2.1 shows all the HRUs considered in this study. Elevations vary from 0 to 1,350 masl with higher altitudes found in the northern part of the country. Tropical wet climate dominates the country however; the south and east experience a tropical monsoon climate. The ranges of maximum and minimum temperatures are from 28–36 °C and 13–25 °C respectively. Temperature varies significantly with location; higher in the plains whereas low in hills. The wet season starts with the monsoon from May to July which extends up to October to November contributing 75 % of total rainfall and consecutively leaving rest part of the year dry with very low available water. Dry period extends longer in the Northeast part of the country even up to May/June. The average annual rainfall of the country is about 1,574 mm which also changes with location. The annual rainfall is about 1,200 mm in the northern mountainous region, 1,300 mm in the central plain, below 1,000 mm in the western strip of the north-east plateau and increases to 1,600 mm towards the Far East end of the north-east plateau. The east coast peninsula receives additional rainfall from the northeast monsoon during November through January and annual rainfalls of 1,800 mm and 2,500 mm are observed over the eastern and western coasts of the peninsula respectively.

2.3 Methodology

2.3.1 Data Collection

2.3.1.1 Hydro-meteorological Data

Daily precipitation data of 95 meteorological stations covering the whole of Thailand were collected for the period of 1971–2010 from Thai Meteorological Department (TMD). The distribution of the numbers of stations from the basins was done based on the area of the HRU and spatial distribution of the stations. Data collected from all stations were used for creating Thiessen polygons for determining the distribution of rainfall in the HRUs. Missing data were generated by creating linear regression models based on observed and gridded daily precipitation dataset from APHRODITE (http://www.chikyu.ac.jp/precip/). The daily river discharge data of all 25 major river basins was collected from Royal Irrigation Department (RID) for a period of 1992–2000. However, in order to set up the model for the nine HRUs the river discharge of the major rivers was used.

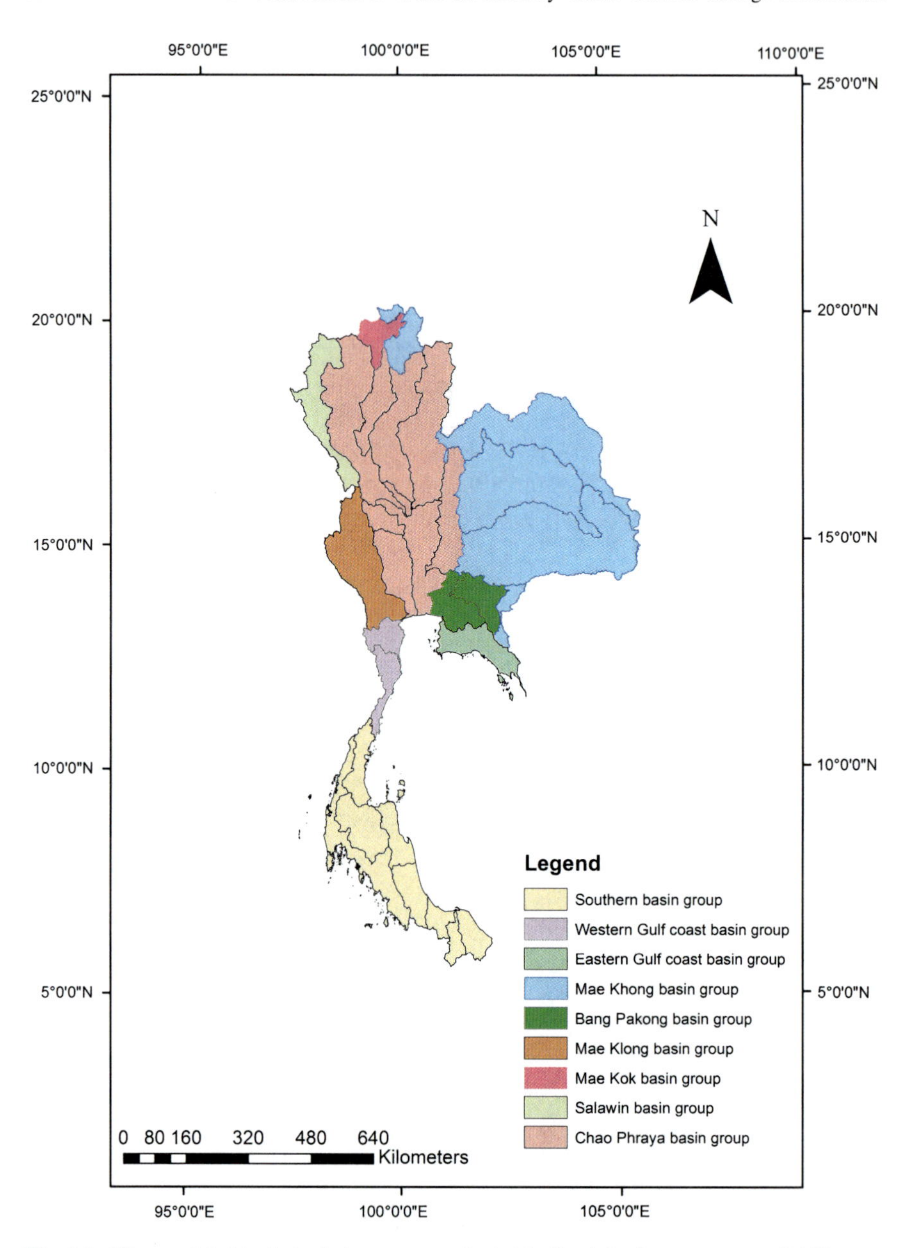

Fig. 2.1 Nine modeled hydrological response units in Thailand for future water availability

2.3.1.2 Future Climate Data

Outputs of the regional climate model (RCM) PRECIS developed by the Hadley Centre of the UK Meteorological Office was used for generating the future gridded climatic dataset. The PRECIS RCM is based on the atmospheric components of the

ECHAM4 GCM from the Max Plank Institute for Meteorology, Germany. The PRECIS data are produced by the Southeast Asian System for Analysis, Research and Training (SEA START) Regional Center for entire Southeast Asian region with a resolution of 0.2° × 0.2° (approximately 22 × 22 km^2). These data comprise of datasets of A2 and B2 emission scenarios from ECHAM4 A2 and B2. The PRECIS data over the periods of 1971–2000 and 2011–2100 for both A2 and B2 scenarios were obtained from SEA START Regional center (http://gis.gms-eoc.org/ClimateChange/index_en.asp).

2.3.1.3 Other Data

The 90 m resolution Digital Elevation Model (DEM) for whole Thailand was downloaded from the Shuttle Radar Topography Mission (SRTM) website: http://glcf.umd.edu/data/srtm/. Soil data and its classification were done based on Food and Agriculture Organization (FAO) recommendations. The land use and land cover map for 2005 was retrieved from Land Development Department (LDD), Government of Thailand.

2.3.2 Methodological Framework

Figure 2.2 represents the framework followed for assessing the impacts of climate change on water availability in Thailand. This study also emphasizes the importance of bias correction for the precipitation data obtained from RCM at basin level. Statistical comparison was done with raw RCM and bias corrected data to evaluate the outputs with the observed data for the current time period. Simultaneously the

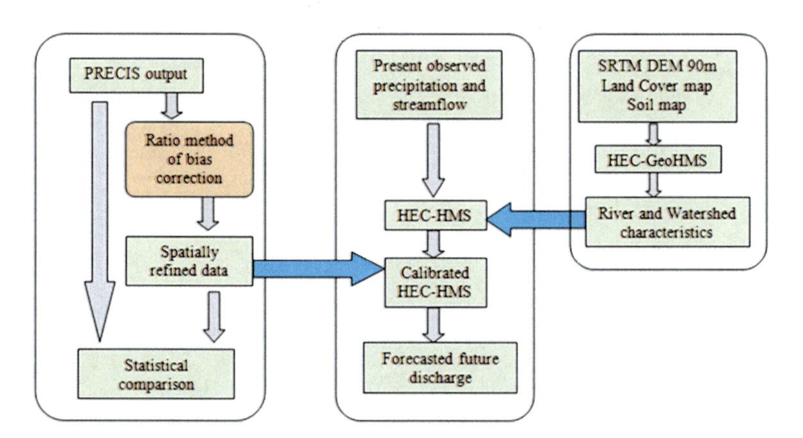

Fig. 2.2 Methodological framework followed to assess the impact of climate change on future water availability in Thailand

semi-distributed hydrological model HEC-HMS version 3.5, developed by United States Army Corps of Engineers was also set up for each HRU by using the outputs of HEC-GeoHMS. The bias corrected precipitation output was then fed into HEC-HMS to simulate the future decadal water availability for A2 and B2 scenarios.

2.3.3 Ratio Method of Bias Correction

The ratio method for bias correction was derived from Braun et al. (2011) which involves three steps. The first deals with determining monthly precipitation over the reference period followed by estimation of the monthly biases by using the mean monthly precipitation and the RCM dataset. Finally the calculation of the fine spatial resolution projected output is calculated based on the observed reference period and monthly bias computed data. Due to handling enormous amount of data for this research; in addition to the satisfactory performance of this method in other basins (Chen et al. 2013; Mpelasoka and Chiew 2009), the particular method was selected. Figure 2.3 illustrates the stepwise flowchart of the bias correction technique.

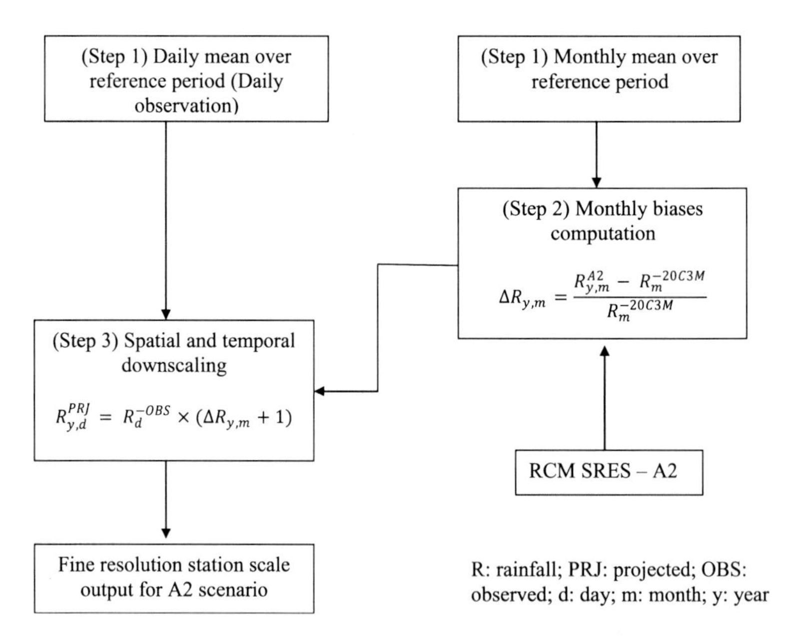

Fig. 2.3 Methodological flowchart for bias correction technique showing A2 scenario (as an example) applied in this study

2.3.4 Hydrological Model

HEC-HMS model was selected to simulate the discharge in all HRUs. HEC-HMS model set up includes the setting of basin model, meteorological model and define the control specifications. The basin parameters including sub-basin area, centroid, slope of basin and longest flow are pre-requisite for HEC-HMS which was further derived by basin delineation from HEC-GeoHMS version 1.1 in ArcView GIS 3.2. The outputs of HEC-GeoHMS for all HRUs were imported to HEC-HMS to set up the model for those particular HRUs. Since the model was run for longer period of time, continuous loss model was chosen. Clark unit hydrograph transformation, constant monthly baseflow and lag routing methods were selected to develop the model for the HRUs. The meteorological model was developed by Thiessen Polygon Weight method. Further details on basin/HRU development can be found in USACE 2000. Observed hydrological and meteorological data for the period of 1995–2003 and 2004–2010 was used for calibration and validation of the model respectively.

2.4 Results and Discussion

2.4.1 Comparison of RCM and Bias Corrected Values

Table 2.1 shows the comparison of the raw RCM and bias corrected rainfall data with the observed for reference period for all nine HRUs. The results clearly indicate that the bias correction gives better results in representing the present day climate. It can be observed that for Chao Phraya and Western Gulf HRUs the raw

Table 2.1 Comparison of RCM simulated and bias corrected average annual precipitation values with the observed precipitation

HRU name	Observed (mm) (2001–2010)	RCM simulated		Bias corrected	
		Absolute value (mm)	% change	Absolute value (mm)	% change
Salawin	1,190 ± 47	1,042 ± 55	−12.46	1,157 ± 34	−2.78
Mae Kok	1,730 ± 63	1,321 ± 78	−23.65	1,645 ± 45	−4.88
Mae Khong	1,944 ± 127	2,122 ± 141	9.16	2,001 ± 114	2.94
Chao Phraya	1,076 ± 162	1,432 ± 154	33.09	1,119 ± 148	4.00
Mae Klong	1,612 ± 27	1,578 ± 45	−2.11	1,602 ± 31	−0.63
Bang Pakong	1,422 ± 44	1,332 ± 64	−6.37	1,401 ± 47	−1.46
Eastern Gulf	1,908 ± 96	1,515 ± 75	−20.61	1,834 ± 91	−3.89
Western Gulf	1,063 ± 89	1,432 ± 78	34.72	1,134 ± 92	6.65
Southern	2,221 ± 124	2,850 ± 98	28.33	2,336 ± 137	5.17

RCM data shows a deviation of +33.09 and +34.72 % in magnitude of observed average annual precipitation whereas removing bias in the dataset can reduce it to +4.00 and +6.65 % respectively. Similar results were also obtained in other HRUs as well however it can also be observed that simulated RCM values depend on the location of the HRU. For instance the percent deviation in simulated precipitation by RCM is higher in the coastal areas whereas in mountains and plains the simulations are in good agreement with the observed values.

2.4.2 Projected Rainfall Anomalies

The rainfall anomalies projected by bias correction of the RCM dataset were calculated for dry and wet seasons separately for 2011–2040 (2020s), 2041–2070 (2050s) and 2071–2099 (2080s). Table 2.2 represents the percent deviation in average rainfall for dry season. It is observed that Mae Kok, Mae Khong and Bang Pakong HRUs will experience higher increase in precipitation in all three future time periods for both scenarios. It can also be observed that Chao Phraya will experience an increase in precipitation up to 22 and 13.5 % for A2 and B2 scenarios respectively in 2080s relative to baseline period. Higher variation in the observed precipitation is also observed for the baseline period. The larger number of stations considered in the HRU can be attributed to this variation.

In case of wet season, Mae Kok and Mae Khong HRUs will experience an increase in precipitation up to 29.5 and 36.5 % in 2080s relative to baseline period (Table 2.3). The elevation of Mae Kok and larger spatial extent of Mae Khong

Table 2.2 Projected rainfall anomalies (%) for dry season in case of A2 and B2 scenarios

Basin group	Baseline (1971–2000)	2020s		2050s		2080s	
	Dry period (Nov–Apr)	A2 (%)	B2 (%)	A2 (%)	B2 (%)	A2 (%)	B2 (%)
Salawin	273.7 ± 33	0.9	−1.6	3.4	4.8	14.3	10.0
Mae Kok	397.9 ± 37	17.1	14.3	21.1	22.1	33.6	28.2
Mae Khong	408.2 ± 94	4.4	14.8	22.7	21.8	37.8	23.7
Chao Phraya	275.5 ± 102	3.1	0.8	7.4	8.4	21.9	13.5
Mae Klong	412.7 ± 21	−4.2	−6.6	−1.9	−0.5	8.4	4.4
Bang Pakong	364.0 ± 24	23.6	23.9	29.3	29.7	32.0	30.6
Eastern Gulf	450.3 ± 51	4.0	1.5	6.5	8.0	17.6	13.3
Western Gulf	318.9 ± 61	−12.9	−15.0	−10.8	−9.6	−1.4	−5.0
Southern	666.3 ± 62	−4.6	−7.4	−7.4	−4.4	−2.4	−3.0

Table 2.3 Projected rainfall anomalies (%) for wet season in case of A2 and B2 scenarios

Basin group	Baseline (1971–2000)	2020s		2050s		2080s	
	Wet period (May–Oct)	A2 (%)	B2 (%)	A2 (%)	B2 (%)	A2 (%)	B2 (%)
Salawin	916.3 ± 52	−0.6	2.8	2.6	2.4	14.3	2.3
Mae Kok	1,332.1 ± 72	12.7	16.8	16.7	15.7	29.5	16.9
Mae Khong	1,535.7 ± 133	14.1	18.4	19.8	20.9	36.5	26.2
Chao Phraya	800.5 ± 181	1.2	4.2	3.7	4.1	16.2	3.3
Mae Klong	1,199.3 ± 43	7.2	10.9	10.7	10.4	23.2	10.3
Bang Pakong	1,057.9 ± 56	11.3	13.5	12.5	14.7	31.9	17.1
Eastern Gulf	1,457.7 ± 102	2.7	6.2	6	5.7	18	5.7
Western Gulf	744.1 ± 118	−3.0	0.3	0.2	−0.1	11.6	−0.1
Southern	1,554.7 ± 132	1.0	5.4	8.8	8.7	21.4	15.4

HRU can be the causative factor for the projected increase. Surprisingly, Mae Klong and Bang Pakong HRUs show higher precipitation in order of 23.2 and 31.9 % for A2 scenario in 2080s. The influence of physical process of the mountains and sea respectively in climatology can be the attributed to this. In addition a higher variation in the observed precipitation for the baseline period is observed for Mae Khong, Chao Phraya, Eastern Gulf, Western Gulf and Southern HRUs.

2.4.3 Calibration and Validation of HEC-HMS

The HEC-HMS was calibrated and validated for all HRUs based on the observed stream flow data. The period of 1995–2003 and 2004–2010 was chosen for model calibration and validation respectively. The modeling results were evaluated based on the coefficient of determination and volumetric error. The results suggest the model estimates the runoff in good agreement with the observed runoff. However, poor relationship is observed for Mae Khong, Eastern Gulf, Western Gulf and the Southern group HRUs (Table 2.4). Multiple outlets in the coastal region can be attributed to the poor performance in these HRUs. However, the performance of the model is still in acceptable range and therefore the projection was carried out for the future time periods.

2.4.4 Projection of Decadal Water Availability at HRU Scale

Figures 2.4 and 2.5 show the simulated future water availability for A2 and B2 scenarios at all HRUs considered. The results suggest that, the future change in

Table 2.4 HEC-HMS model performance statistics during calibration and validation

HRU name	Coefficient of determination (R^2)		Volumetric error (%)	
	Calibration	Validation	Calibration	Validation
Salawin	0.78	0.76	3.87	2.56
Mae Kok	0.74	0.71	4.23	3.12
Mae Khong	0.61	0.67	8.74	6.58
Chao Phraya	0.87	0.82	1.42	2.06
Mae Klong	0.85	0.81	2.64	3.51
Bang Pakong	0.77	0.84	4.11	3.28
Eastern Gulf	0.62	0.59	−9.84	−10.71
Western Gulf	0.70	0.71	−9.71	−8.65
Southern	0.65	0.62	−8.21	−8.65

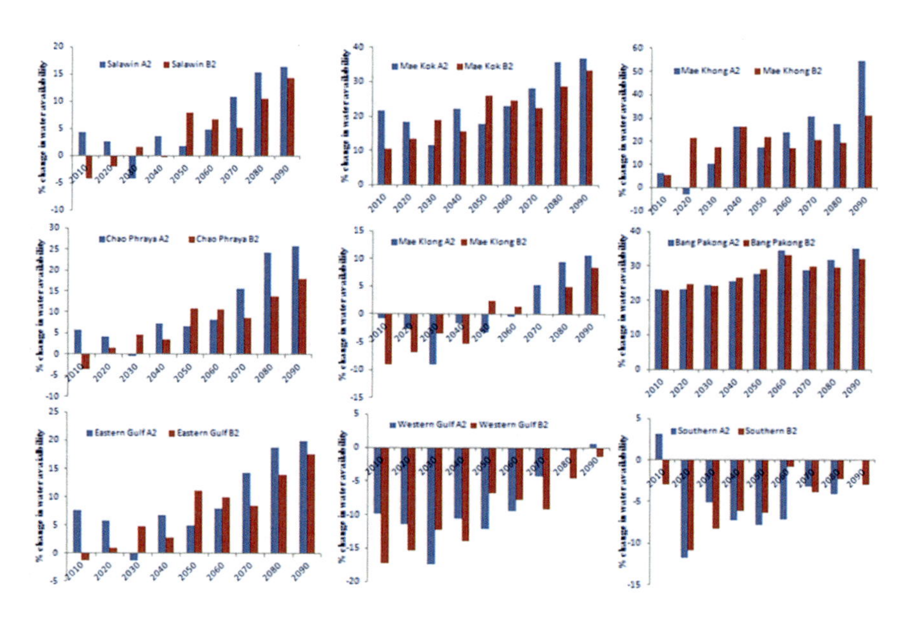

Fig. 2.4 Percent change in decadal water availability for dry season in all HRUs for A2 and B2 scenarios

water availability is univocal in some HRUs whereas it is in contradiction in others. For instance, in case of Salawin, Chao Phraya and Eastern gulf HRUs for dry season, the water availability is expected to fluctuate in the first three decades followed by an increasing trend for both scenarios. It can also be noted that, the magnitude of available water is higher for A2 scenario relative to B2. It can also be observed that for the corresponding season at Mae Kok, Mae Khong and Bang

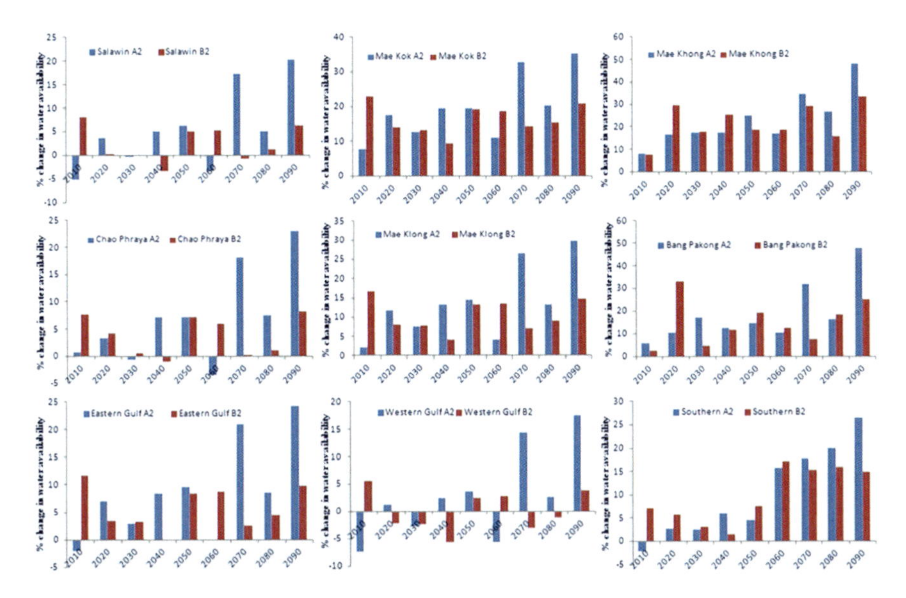

Fig. 2.5 Percent change in decadal water availability for wet season in all HRUs for A2 and B2 scenario

Pakong HRUs an increasing trend in available water persists for both scenarios. A contradictory decreasing trend is expected for the coastal HRUs particularly Western and Southern HRUs. However simulation suggests that in Mae Klong HRU, the water availability will be reduced in the first half followed by an increase at the latter half of century. The projected rainfall in those HRUs can be attributed to the pattern of the simulated runoff. Simulation also suggests the southern and western basin groups are expected to experience a decline in the water availability for early part of the century up to 17 and 12 % for 2010s and 2020s respectively relative to present water availability. The projected spatial variability in the water availability due to climate change may significantly affect the long term water management plans.

Simulation of future water availability for wet season suggests a relatively lesser altercation for many HRUs in the country. An increasing trend in water availability is observed for Mae Kok, Mae Khong, Mae Klong, Bang Pakong and the Southern HRUs where the stream flow is expected to increase for all the decades relative to the baseline period for both scenarios. The Salawin, Chao Phraya and Western Gulf shows higher fluctuation in water availability for the decades however; the water availability is expected to increase up to 21, 25 and 17 % for the respective HRUs for 2080s in case of A2 scenario leading to higher focus on increased intensity of flood. Eastern Gulf HRU shows a positive fluctuation in the water availability although an increase up to 23 % is expected for 2080s in case of B2 scenario.

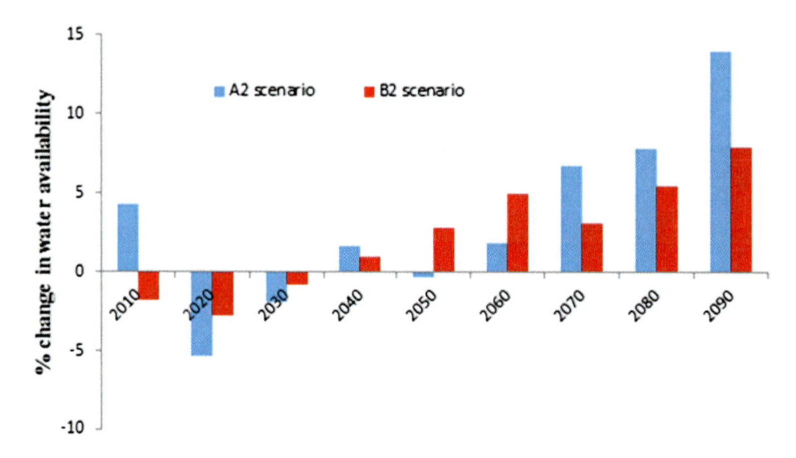

Fig. 2.6 Projected percent change in decadal water availability at national level for A2 and B2 scenarios in case of dry season

Nevertheless, for all HRUs an increasing trend or minor positive fluctuation in water availability is expected for all decades and the possibility of low intensity and frequent foods in wet season in the region.

2.4.5 Projection of Decadal Water Availability at National Scale

Figure 2.6 indicates the national level water availability for the future climate. It can be observed that majority of the HRUs indicates an increasing trend of water availability in dry season, yet at national scale the water availability results demonstrates a dropdown up to 6 % for A2 scenario in 2020s compared to the baseline period of 77,061 MCM. The simulation also illustrates ahead of 2040s, an increase in magnitude is expected with maximum value at 13 and 7.5 % for A2 and B2 scenarios in 2080s. The reduced stream flow in the early decades for most of HRUs attributed to climate change is the influencing factor for the reduced water availability for corresponding time intervals. The results indicate that at national level proper plans for energy and other associated sectors are necessary to be evaluated since in the early decades the water availability is expected to decrease.

In contrast to dry season, wet season water availability is expected to have an increasing trend at national level irrespective of basin scale projection (Fig. 2.7). An increase in water availability of 31 and 17 % are expected for 2080s in case of A2 and B2 scenarios respectively at national scale compared to 183,050 MCM for baseline period which calls for improved land use planning. It can also be noted that the national level water availability is influenced by the size of the HRU considered. The severity in magnitude of precipitation for A2 scenario is the contributing factor

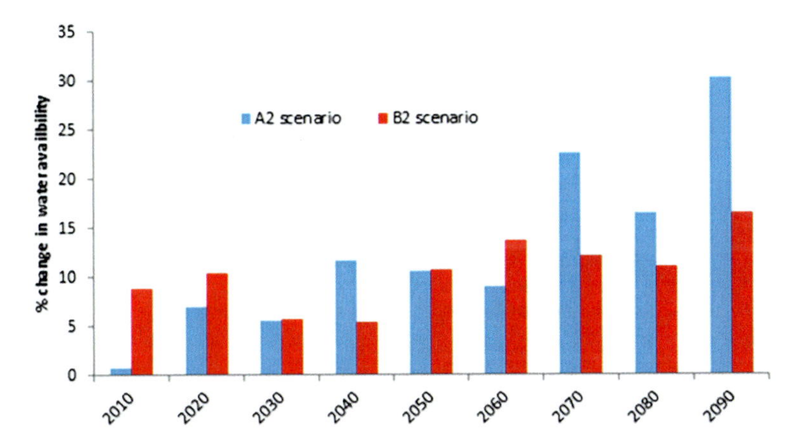

Fig. 2.7 Projected percent change in decadal water availability at national level for A2 and B2 scenarios in case of dry season

for the higher stream flow generation. For instance the stream flow generated for dry season follows similar trend as that of Chao Phraya and Mae Khong HRUs. Similarly for wet season the water availability trend follows similar as that of Mae Klong and Southern basin group HRUs.

2.5 Conclusions

The present study examines the future water availability for Thailand grouped at different HRU and national scale under two different climate change scenarios. The outputs of PRECIS RCM were selected to construct the climate change scenarios for the study area. A comparison of raw RCM outputs and bias correction results suggest the future climate data can be significantly corrected by ratio method of bias correction. Further, bias correction of results illustrates Mae Kok, Bang Pakong, Mae Khong and Southern basin HRUs are expected to have an increase in precipitation ranging from 21.4 to 37.8 % and 15.4 to 30.6 % for A2 and B2 scenarios respectively considering both dry and wet seasons by 2080s. Hydrological model simulation suggests that for both scenarios and seasons; all HRUs show similar trend except for Mae Klong, Bang Pakong, Western and Southern HRUs where dry season indicates different trend relative to that of wet period. Although, in all cases the extreme water availability is observed in 2080s ranging from −7 to 47 % and −17.8 to 54 % for wet and dry periods respectively relative to baseline period. The national level water availability varies from −5.5 % in 2020s to +13 % 2090s and +1 % in 2010s to +29 % in 2080s. The increasing trend of water availability indicates better water management plan for the increased risks of flood in the nation.

References

Arnell NW (2003) Effect of IPCC SRES emission scenarios on river runoff: a global perspective. Hydrol Earth Syst Sci 7(5):619–641
Babel MS, Bhusal SP, Wahid SM, Agarwal A (2013) Climate change and water resources in the Bagmati River Basin, Nepal. Theoret Appl Climatol. doi:10.1007/s00704-013-0910-4
Braun M, Caya D, Frigon A, Slivitzky M (2011) Internal variability of the Canadian RCM's hydrological variables at the basin scale in Quebec and Labrador. J Hydrometeorol 13:443–462
Boyer C, Chaumont D, Chartier I, Roy AG (2010) Impact of climate change on the hydrology of St. Lawrence tributaries. J Hydrol 384:65–83
Chen J, Brissette FP, Chaumont D, Braun M (2013) Performance and uncertainty evaluation of empirical downscaling methods in quantifying the climate change impacts on hydrology over two North American river basins. J Hydrol 479:200–214
Chiew FHS, Teng J, Vaze J, Kirini DGC (2009) Influence of global climate model selection on runoff impact assessment. J Hydrol 379:172–180
Chu X, Steinman A (2009) Event and continuous hydrologic modeling with HEC-HMS. J Irrig Drain Eng. January/February 119
Davis EGR, Simonovic SP (2011) Global water resources modeling with an integrated model of the social – economic – environmental system. Adv Water Resour 34(6):684–700
Dore MHI (2005) Climate change and changes in global precipitation patterns: what do we know? Environ Int 31:1167–1181
Graiprab P, Pongput K, Tangtham N, Gassman PW (2010) Hydrologic evaluation and effect of climate change on the at Samat watershed, Northeastern Region, Thailand. Int Agric Eng J 19 (2):12–22
Intergovernmental Panel on Climate Change (IPCC) (2001) McCarthy JJ, Canziani OF, Leary NA, Dokken DJ, White KS Climate change 2001: impacts, adaptation and vulnerability. Contribution of working group II to the third assessment report of the intergovernmental panel on climate change, Cambridge University Press, Cambridge
Intergovernmental Panel on Climate Change (IPCC) (2007) Summary for policy makers. In: Parry ML, Canzani OF, Palutikof JP et al. Climate change 2007: impacts, adaptation and vulnerability. Contribution of working group II to the fourth assessment report of the intergovernmental panel on climate change, Cambridge University Press, Cambridge
Jiang T, Su B, Hartmann H (2007a) Temporal and spatial trends of precipitation and river flow in the Yangtze River Basin, 1961–2000. Geomorphology 85:143–154
Jiang T, Chen YD, Xu C, Chen X, Chen X, Singh VP (2007b) Comparison of hydrological impacts of climate change simulated by six hydrological models in the Dongjiang Basin, South China. J Hydrol 336:316–333
Koutroulis AG, Tsanis IK, Daliakopoulos IN, Jacob D (2013) Impact of climate change on water resources status: a case study for Crete Island, Greece. J Hydrol 479:146–158
Kranz N, Menniken T, Hinkel J (2010) Climate change adaptation strategies in the Mekong and Orange-Senqu basins: what determines the state of play? Environ Sci Policy 13:648–659
Leander R, Buishand TA (2007) Resampling of regional climate model output for the simulation of extreme river flows. J Hydrol 332:487–496
Li F, Zhang Y, Xu Z, Teng J, Liu C, Liu W, Mpelasoka F (2013) The impact of climate change on runoff in the southeastern Tibetan Plateau. J Hydrol 505:188–201
Minville M, Brissette F, Leconte R (2008) Uncertainty of the impacts of climate change on the hydrology of a Nordic watershed. J Hydrol 358:70–83
Mpelasoka FS, Chiew FHS (2009) Influence of rainfall scenario construction methods on runoff projections. J Hydrometeorol 10:1168–1183
Nohara D, Kitoh A, Hosaka M, Oki T (2006) Impact of climate change on river runoff. J Hydrometeorol 7:1076–1089
SEA START RC (2009) Water and climate change in the lower Mekong basin: diagnosis and recommendations for adaptation. Water and development research group. Helsinki University

of Technology (TKK) and Southeast Asia START Regional Center, Chulalongkorn University. Water and Development Publications, Helsinki University of Technology, Espoo, Finland

Sharma D, Babel MS (2013) Application of downscaled precipitation for hydrological climate-change impact assessment in the upper Ping River Basin of Thailand. Clim Dyn. doi:10.1007/s00382-013-1788-7

Sharma D, Gupta AD, Babel MS (2007) Spatial disaggregation of bias-corrected GCM precipitation for improved hydrologic simulation: Ping River Basin, Thailand. Hydrol Earth Syst Sci 4:35–74

USACE (2000) Hydrological modeling system HEC-HMS: technical reference manual. US Army Corps of Engineers, Hydrologic Engineering Center, Davis

Weart S (2010) The development of general circulation models of climate. Stud Hist Philos Mod Phys 41:208–217

Chapter 3
Climate Change Impact on Reservoir Inflows of Ubolratana Dam in Thailand

Abstract This study analyzes the future climate implications on the reservoir inflows for Ubolratana dam, Thailand. The future climate data of precipitation, maximum and minimum temperature was derived from regional climate model RCM Providing Regional Climates for Impact Studies (PRECIS) for A2 and B2 climate scenarios. Bias correction was performed on the climate data for finer spatial resolution. Future inflow was estimated by the simulation of the future flow by hydrological model MIKE 11 NAM module. The results suggest elevated maximum and minimum temperatures relative to the baseline period. Similarly, higher intense rainfall is also expected in the future for both scenarios considered. Hydrological model simulation results for future climate suggests higher inflows in the future for both scenarios however less intense in case of B2 scenario. Resilience, reliability and vulnerability (RRV) analysis show that with the increasing rainfall in future will contribute to lower resilience and reliability whereas higher vulnerability.

3.1 Introduction

Thailand's major economy relies on agriculture and it is expected to be seriously affected by the adverse impacts of climate change due to the high sensitivity of agriculture on climatic variables including water resources. Studies indicate among Southeast Asia, Thailand has the highest per capita water use and 94 % to total water use is accounted for agricultural sector (Chulalongkorn 2012). The vulnerability of freshwater resources attributed to climate change is undoubtedly negligible in the region since the development of the region depends on water resources. Moreover, the observed changes in water regime driven by climatic factors have not only affected agriculture but also the energy production in past decades (Hunukumbura and Tachikawa 2012).

Although not many researches focusing on climate change has been conducted in Chi River basin of Thailand however; the existing findings illustrate the presence

S. Shrestha, *Climate Change Impacts and Adaptation in Water Resources and Water Use Sectors*, Springer Water, DOI 10.1007/978-3-319-09746-6_3

of ambiguous and increasing trends in precipitation and temperature respectively (Artlert et al. 2013). An increase of 1.2 to 1.9 °C in temperature is projected by 2050s relative to historical period under climate change in Thailand (IPCC 2007). Further studies on climate change has illustrated that the shifts in precipitation pattern are not coherent and therefore it has its implications on regional scale (Manomaiphiboon et al. 2013). The observed and projected changes in the climatic variables will have significant influence on the streamflow and watershed hydrology (Sivakumar 2011). The alteration of the rainfall pattern will certainly influence in the seasonal reservoir inflows and therefore shift in the reservoir operations are necessary. Although on a global scale majority of studies have mainly focused on the downstream beneficial interests in large river systems; merely a handful of studies have focused on the climate change implications on the reservoir inflows (Lauri et al. 2012; Raje and Mujumdar 2010).

General Circulation Models (GCMs) are tools which provides the future climatic data for a given greenhouse emission scenario. However due to the coarse spatial resolution, it is not suitable to apply the outputs at basin scale or sub-grid level hydrological assessment studies. Statistical or dynamic downscaling [regional climate models (RCMs)] methods are generally applied for refining the climatic data for catchment modeling. Even though some studies even have applied RCM data directly in impact assessment studies yet globally, in many basins output of 20–50 km resolution are not sufficient to represent the true climate of the regions at station level and hence further downscaling or bias correction are suggested (Sharma and Babel 2013; Teutschbein and Seibert 2012).

In order to complement these problems, the present study was conducted to forecast the future reservoir inflows using bias corrected future climate data and hydrological model for IPCC special report on emission scenarios (SRES) A2 and B2 for Ubolratana dam located in Chi river basin, Northeast Thailand. The main objectives are: (1) to investigate the future climate change in the upstream of the Ubolratana reservoir and (2) response of the climate change on the reservoir inflows for future periods.

3.2 Study Area

The Chi River Basin is located in the north-east of Thailand extending from 15° 30′–17° 30′ N latitude and 101° 30′–104° 30′ E longitude and covers an area of 49,477 km^2 in twelve provinces extending about 360 km from the east to west and 210 km from the north to south. Figure 3.1 shows the location of the study area in the basin. The climate is moderately tropical with average annual temperature ranging from 26.6 to 27.8 °C. Further, the region is dominated by two monsoon seasons namely the southwest which influences from mid-May to mid-October with heavy showers and the northeast monsoon extending from mid-November to mid-February which accumulates to 1,700 mm of average annual rainfall. South China Sea contributes to the tropical cyclone in the region from August to September.

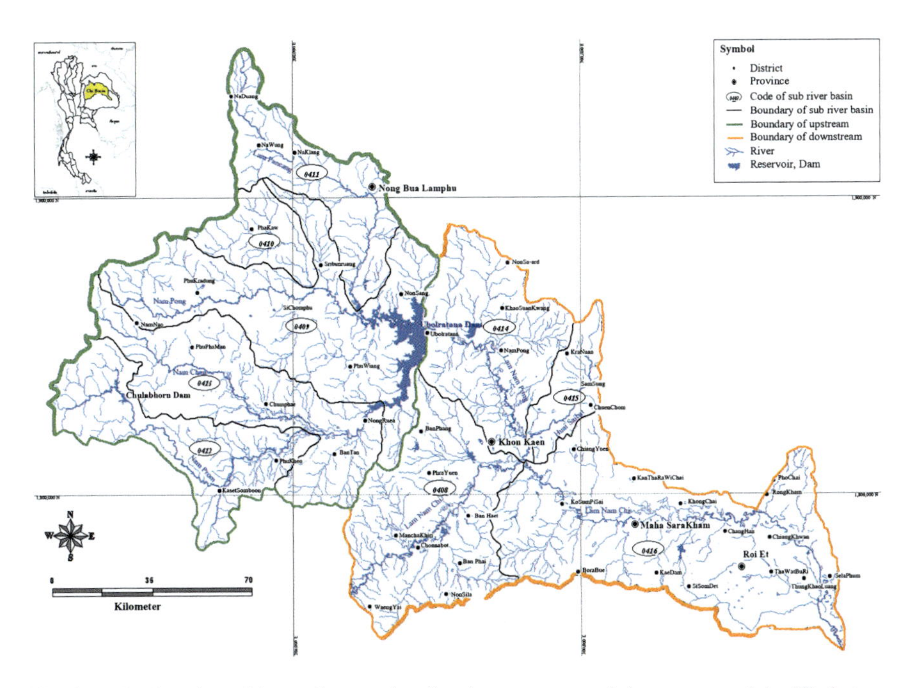

Fig. 3.1 The location of the study area showing the upstream and downstream of the Ubolratana dam

Long term observation suggests the average annual runoff of the Chi river basin is 11,244 MCM which can be disintegrated into 9,638 MCM for wet and 1,606 MCM for dry season (RID 2002).

3.2.1 Characteristics of Ubolratana Dam

The Ubolratana is a multipurpose dam with a catchment area of 12,000 km^2 for development of electricity, irrigation, flood control, transportation, fisheries and tourism. The dam is located on the Nam Pong River at Kok Soong Sub-district, Ubolratana District of Khonkaen province. The structural is an earth core rock-fill dam and was constructed in 1984 with the height of 32 m, crest length of 885 m, crest width of 6 m. Normal Flood Level is 182.00 m (MSL) with a maximum storage capacity is 2,559 MCM. The total catchment area is 12,000 km^2.

3.2.2 Flood Problems in Ubolratana Reservoir

Historical data suggests the average inflow in the reservoir is 2,481 MCM which is equivalent to the capacity of the reservoir and therefore during the extreme rainfall

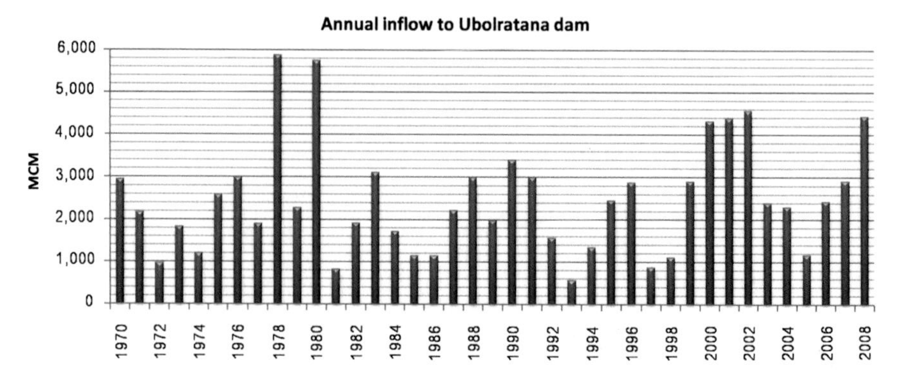

Fig. 3.2 Record of annual inflow from 1970 to 2008 to Ubolratana dam

years the water resource management is a big issue in the reservoir. Figure 3.2 illustrates the annual inflow from 1970 to 2008 in the dam. It can be observed in 1978 and 1980 the inflow was twice the capacity and therefore due to the safety concerns spillways had to operate far beyond the designed discharge which led to flash flood in downstream.

3.2.3 Data Collection

Two sets of meteorological data (rainfall and temperature) in the Pong river sub-basin were collected for 81 stations lying within and adjacent to the basin according to data availability and frequency. In addition streamflow data from 26 stations were extracted for the upstream and downstream of the reservoir. The meteorological data was obtained from Thai Meteorological Department (TMD) whereas, the streamflow data was retrieved from Royal Irrigation Department (RID).

The future climate data were retrieved from the RCM Providing REgional Climates for Impact Studies (PRECIS) which is developed by Hadley Centre at UK Met Office (http://cc.start.or.th/). The model has a spatial resolution of 20 km and derives its boundary conditions from the GCM ECHAM4. Sharma et al. (2007) evaluated suitability of several GCMs in Ping river basin (an adjacent basin); findings suggest ECHAM4 is the best suitable model in the basin in order to represent the precipitation and temperature for the historical climate.

3.3 Methodology

Figure 3.3 illustrates the methodological flowchart applied in this study. First of all the bias correction of the PRECIS dataset for A2 and B2 scenarios were done at all respective stations contributing to the basin which was followed by set-up of the

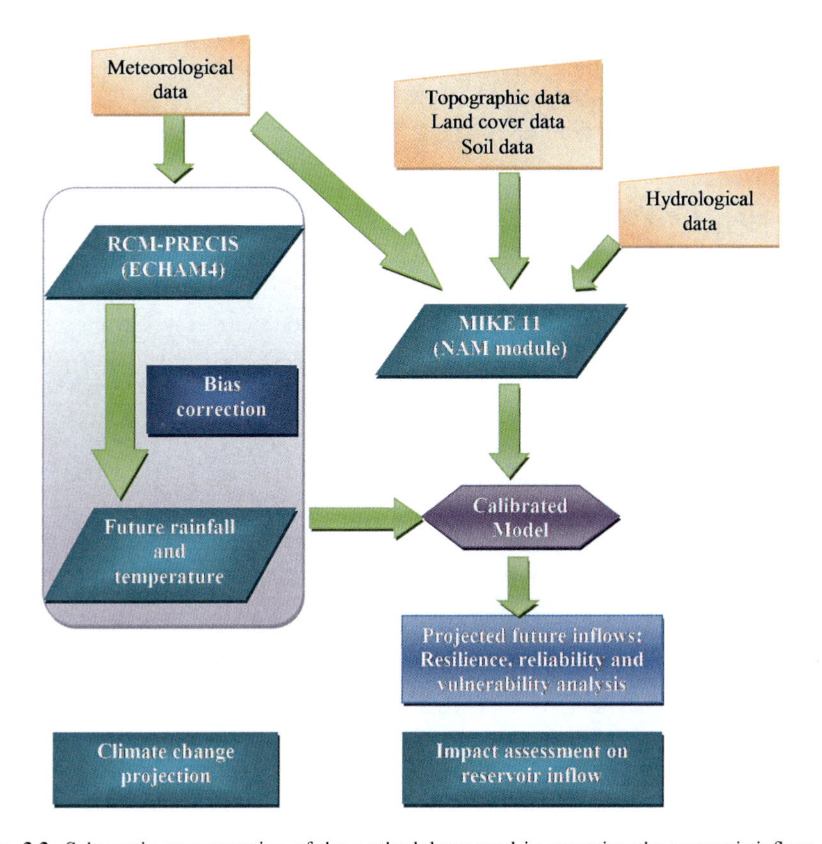

Fig. 3.3 Schematic representation of the methodology used in assessing the reservoir inflow

MIKE 11 rainfall-runoff model (NAM) for the upstream of reservoir. Further, the calibrated and validated hydrological model was used for projection of future reservoir inflows under the considered climate scenarios. Additionally resilience, reliability and vulnerability analysis was also done for the future inflow.

3.3.1 Extraction of Climate Projection Data

Although there were 81 stations considered in the study however; based on continuity of available dataset 39 rain gauge stations were selected to create the Thiessen interpolation. The selection scheme of rain gauge was done based on the distance of each station, the completion and reliability of rainfall data. Existing gaps in the raingauge station data were interpolated by using square inverse distance interpolation method from nearby stations which had equivalent rainfall. Moreover, RCM provides the rainfall data in the form of grids (20 × 20 km) at its center. Data from 42 grids were retrieved which were covering all the upstream and downstream raingauge stations. In order to get better estimation of the rainfall from the RCM data weights were determined for each station based on the fraction of the Thiessen polygon area lying on each RCM grid.

3.3.2 Bias Correction

Due to the spatial dependence of the biases in temperature and precipitation, performing bias correction is necessary for each station. The biases from the temperature were removed by power law transformation theorem where the data is normally distributed. It uses the scaling and shifting of the mean and variance of the dataset (Leander and Buishand 2007). Further the bias correction for precipitation was also done by non-liner method of multi-day window for correction of coefficient of variation (CV). The baseline period considered for correcting the future period dataset was 1976–2005. Additional details on bias correction used for this study can be found on Leander and Buishand (2007). Future climate projections were done for three time windows 2010–2039 (2020s), 2040–2069 (2050s) and 2070–2099 (2080s).

For correcting rainfall, although the block length for bias correction as recommended by Leander and Buishand (2007) is 5 days; however, Terink et al. (2010) suggested if the block lengths are chosen too small, high chances of correcting the natural variability rather than correcting the systematic model error exists. Based on this recommendation, in the present study, a sensitivity analysis for different block lengths 15, 25, 35, 45 and 65 days was done to represent the best performance of statistic for bias correction. Moreover, the performance of the multi-day analysis was assessed by calculating Root Mean Square Error (RMSE) and Efficiency Index (EI) at monthly scale.

3.3.3 Hydrological Model

One dimensional hydrodynamic model MIKE 11 was applied to simulate the flow and the water level of the river. The computational core of MIKE 11 is hydrodynamic simulation engine and complements a wide range of additional modules and extensions. The general rainfall-runoff modules integrated in this model are the Nedbør-Afstrømnings-Model or NAM, the unit hydrograph method (UHM), conceptual continuous hydrological model, a monthly soil moisture accounting model, runoff methods tailored to urban environments (URBAN) and semi-distributed rainfall-runoff-geomorphological approach (DRiFt). For our present study NAM approach was used due to its suitability for large river basins with numerous catchments with complex networks of rivers (DHI 2007).

The MIKE 11/NAM model represents the various components of the rainfall–runoff process by continuously accounting for the water content in four different

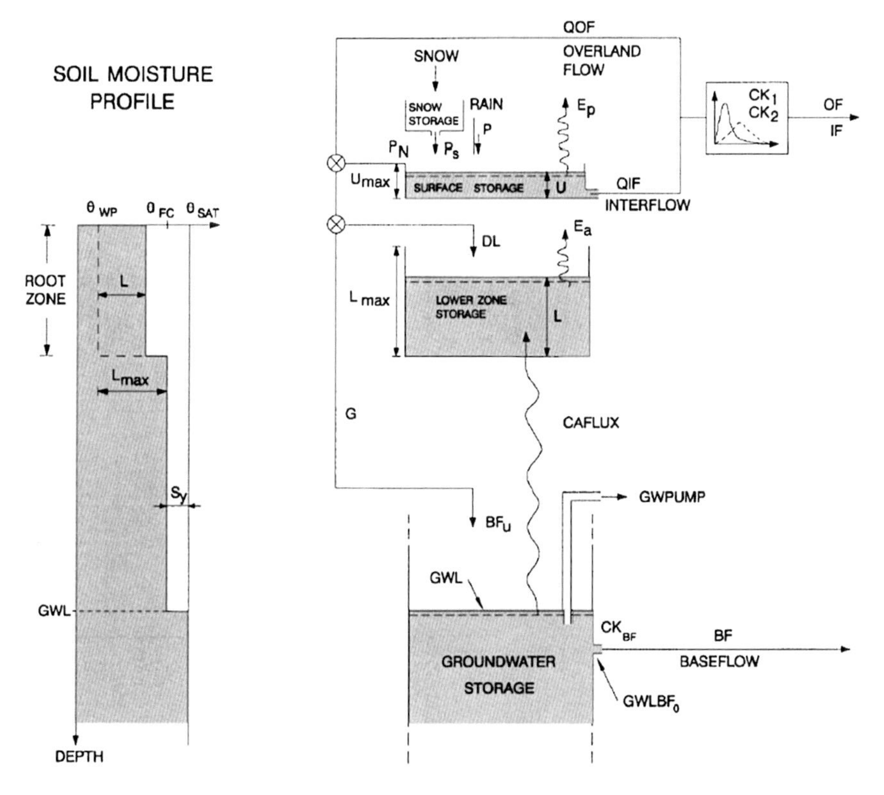

Fig. 3.4 NAM model structure

and mutually interrelated storages namely snow, surface, lower zone and groundwater storage where the each storage represents different physical elements of the catchment (Fig. 3.4). The basic input requirements for the NAM model consist of model parameters, initial conditions, meteorological data and stream flow data. In the present application, the nine most important parameters of the NAM model are determined by calibration.

Split-sample testing scheme was used for validation of the model which suggests calibration of a model based on 3–5 years of data and validation for another period of similar length (Klemeš 1986). Calibration period from 2003 to 2007 was chosen to represent the recent climate whereas validation period was selected from 1998 to 2002. The selection of the calibration and validation periods considers all the low and extreme flows and therefore better model set up was expected. The future inflow projections were done for the bias corrected three time windows 2020s, 2050s and 2080s. The changes in the future inflow were analyzed based on the average monthly flows and daily flow duration curves.

In addition to the projection of the change in inflows under climate change, reliability (C_R), resilience (C_{RS}) and vulnerability (C_V) (RRV) analysis was also done in order to evaluate the performance of the inflows. RRV criteria were evaluated under future climate conditions both A2 and B2 emission scenarios. First a criterion, C, is defined for the normal range of inflow, where an unsatisfactory condition occurs when inflow is out of normal range. According to Fowler et al. (2003), the normal range of inflow was between 20th and 80th percentile of historical inflow data (1970–2008). If the inflow is in normal range, we can conclude to be in a satisfactory (S) state, otherwise in an unsatisfactory (U) state (Eq. 3.1).

$$Z_t = \begin{cases} 1, & \text{if } X_t \in S \\ 0, & \text{if } X_t \in U \end{cases} \tag{3.1}$$

Where: Z_t is a generic indicator of satisfactory or unsatisfactory. Another indicator, W_t, which represents a transition from S to U states, is defined as in Eq. 3.2

$$W_t = \begin{cases} 1, & \text{if } X_t \in U \text{ and } X_t + 1 \in S \\ 0, & \text{otherwise} \end{cases} \tag{3.2}$$

Furthermore, if the periods of X_t is in unsatisfactory state then based on $J_1, \ldots, J_N$ where N is the number of U periods, then reliability, resilience and vulnerability indices during the total time period (T) was calculated as in Eqs. 3.3, 3.4 and 3.5 respectively.

Reliability

$$C_R = \frac{\sum_{t=1}^{T} Z_t}{T} \tag{3.3}$$

Resilience

$$C_{RS} = \frac{\sum_{t=1}^{T} W_t}{T - \sum_{t=1}^{T} Z_t} \tag{3.4}$$

Vulnerability time

$$Cv = \max\{J1, J2, \ldots JN\} \tag{3.5}$$

3.4 Results and Discussion

3.4.1 Performance of Bias Correction for RCM Data

The RCM outputs forced by ECHAM5 were bias corrected by applying the power law transforms for rainfall data and the linear approach for temperature data. The

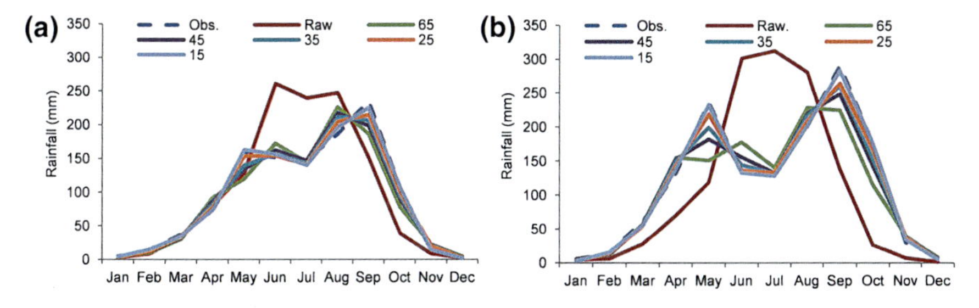

Fig. 3.5 Comparison of multi-day blocks for two RCM grids (a) Id 1681024 and (b) Id 1661016 at the study site

observed data for the 30-year period of 1976–2005 were used as a baseline in this study due to available climate data.

The comparison of the observed and raw RCM data along with multi-day analysis for the baseline period and two grids of RCM from the study area are shown in Fig. 3.5a and b. From visual observation, all multi-day blocks considered are found to follow the similar pattern of monthly rainfall for the grid Id 1681024 except the raw data which is observed to deviate widely relative to the observed values. Further, for grid Id 1661016, a significant deviation in rainfall is observed in May to August where the raw RCM data overestimates the rainfall significantly. However, the multi-day data analysis estimates suggest larger blocks leads to greater variation especially in the months with higher rainfall. Moreover, low day blocks (15 and 25 days) visually perform well in line with the observed values.

Performance indicators (RMSE and EI) calculated for the grids suggest 25 days block is the most suitable since it represents the least RMSE and highest EI relative to the other blocks considered (Table 3.1). As expected, lowest performance is

Table 3.1 Performance statistics of multi-day analysis for two selected grids at the study site for the observed climate

Performance indices	Block lengths	Grid Id 1681024	Grid Id 1661016
RMSE (mm)	Raw (days)	75.68	106.14
	65	69.31	72.17
	45	71.57	64.13
	35	55.13	40.32
	25	36.14	37.66
	15	40.16	37.98
EI	Raw (days)	−45.30	−23.17
	65	−1.29	−8.92
	45	−0.64	−6.67
	35	−0.17	0.16
	25	0.62	0.73
	15	0.16	0.55

observed for the raw data with highest RMSE of 75.68 and 106.14 mm and poorest EI of −45.3 and −23.17 for the two grids. In addition, a relative low performance is also observed for the lowest block (15 days) compared to 25 days block. The correction of the natural climatic variability due to small block size may the attributing factor for the observed low performance. On contrary, 5 day blocks were considered for maximum and minimum temperature. The comparisons of the bias corrected results suggest good capability of the representativeness of the observed temperature at all stations considered.

3.4.2 *Projection of Rainfall*

Projected rainfall under climate change for both scenarios indicates higher intensity of rainfall for all time windows relative to the historical climate (Fig. 3.6). Moreover, it is also evident, for the past climate observed average annual rainfall is 1,900 mm however, by 2080s average annual rainfall is expected to elevate up to 3,000 mm for both scenarios. Furthermore, a shift in the probability of being less precipitation is observed which is highest for A2 scenario relative to B2 which increase higher risks of floods. In addition, from the annual rainfall analysis it is clear that under both scenarios, the magnitude of the mean, median and the quintiles of rainfall are expected to elevate in the future from 1,200 to 1,600 mm in the last part of the century for A2 scenario and 1,650 mm as per the projection for B2 scenario (Fig. 3.7a and b). However, the median values of annual rainfall are 1,550 and 1,400 mm respectively for the corresponding scenarios.

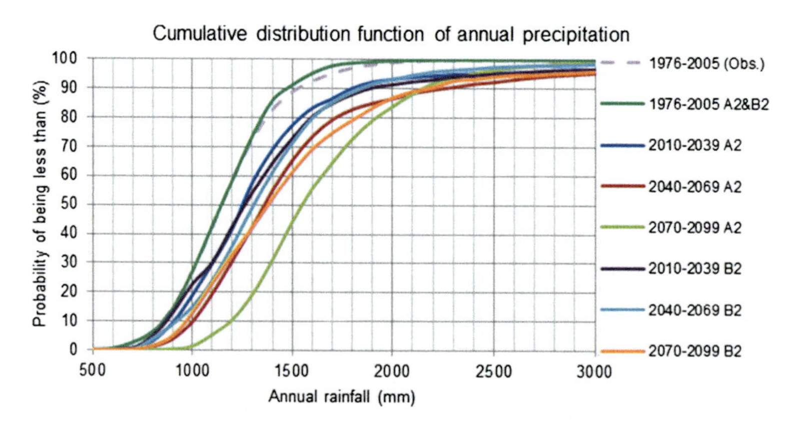

Fig. 3.6 Cumulative distribution function of projected annual rainfall for each time windows considered under A2 and B2 scenario at upstream of Ubolratana dam

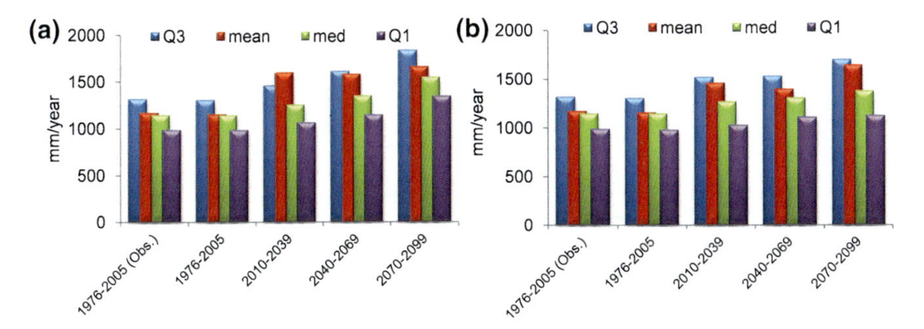

Fig. 3.7 Projected annual rainfall for each period under SRES (a) A2 and (b) B2 at upstream and downstream area of the Ubolratana dam

3.4.3 Projection of Temperature

Bias correction for maximum and minimum temperature suggests an increase in magnitude in future (Fig. 3.8a and b). The highest increase is observed in case of the late century for both scenarios with A2 responds to be severe. Analysis on the change of the maximum and minimum temperature reveals both follows similar trend of shift (Fig. 3.9a and b). The minimum change is observed in case of May whereas; maximum is predominant in July for all the scenarios and time widows considered. Interestingly a significant decline the change is observed for the November and December relative to other months although the magnitude of change is higher relative to May. Nevertheless, it can be summarized that the maximum and minimum temperatures for the basin is expected to increase for all the time periods and scenarios in the future with maximum shift in the A2 scenario and July.

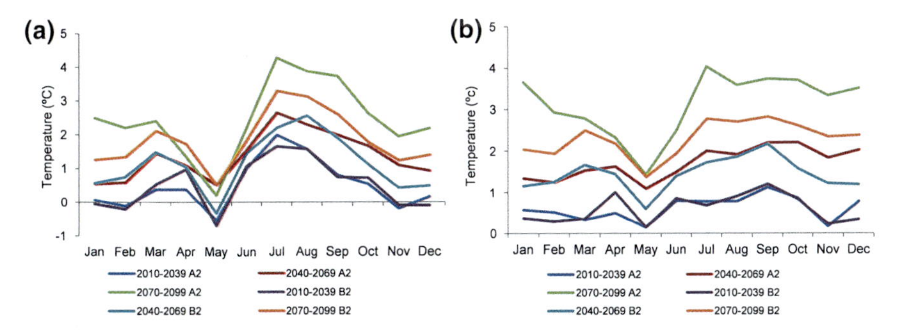

Fig. 3.8 Projected mean monthly (a) maximum and (b) minimum temperature under A2 and B2 scenarios for future time windows at the study site

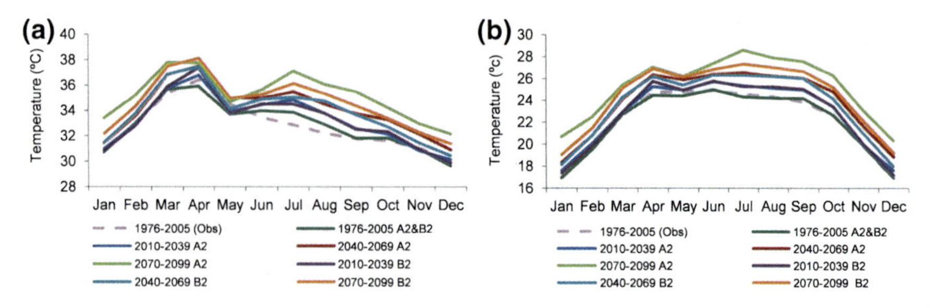

Fig. 3.9 Projected changes in the mean monthly maximum and minimum temperatures for A2 and B2 scenarios for future time periods in study site

3.4.4 Runoff Modelling

3.4.4.1 Model Setup

The hydrological model NAM was calibrated at daily time step with the fine tuning of the parameters as presented in Table 3.2. The model was calibrated by iterating the simulation by changing values of one parameter within the range provided in Table 3.2 and keeping other parameter values constant. Comparison of the simulated and observed discharge in terms of various model evaluation indexes validates the model can simulate the runoff in good agreement with the observed values in the basin (Table 3.3). Although, higher volumetric error (−11.304 %) is observed in case of validation which is probably due to the inability of the model to capture the extreme high flow observed in case of 2002. Also, higher Efficiency Index (EI) and Coefficient of Determination (R^2) is observed for both calibration and validation reflecting the applicability of the model in the study site.

3.4.4.2 Future Runoff Projection

The comparison of the mean monthly inflow to the reservoir for the historical period and the future suggests an increase in the magnitude of the inflow for future under both scenarios considered (Fig. 3.10). In addition, an insignificant shift in the peak is also noticeable for all the future time windows relative to the historical period. Surprisingly, in case of A2 scenario for 2020s, double peak is observed the first in February and second in August. The maximum peak flow (18,000 m^3/s) can be observed for 2080s under A2 scenario whereas a relative lower magnitude of peak flow (13,700 m^3/s) is observed for the corresponding time period for B2 scenario. Furthermore, a significant increase in the peak flow is also observed for the 2050s time window under both scenarios. The expected increase in the flow under future climate indicates higher intensity of flood under future climate.

The flow duration curve generated based on the simulation results suggests the percentage of time that inflow to the dam is likely to equal or exceed some specified

Table 3.2 NAM model parameters calibrated for the basin

Parameter	Description	Lower limit	Upper limit	Calibrated value
U_{max} (mm)	Maximum water content in the surface storage. This storage can be interpreted as including the water content in the interception storage, in surface depression storages, and in the uppermost few cm's of the soil	0	35	20
L_{max} (mm)	Maximum water content in the lower zone storage. L_{max} can be interpreted as the maximum soil water content in the root zone available for the vegetative transpiration	50	350	300
CQOF (–)	Overland flow runoff coefficient. CQOF determines the distribution of excess rainfall into overland flow and infiltration	0	1	0.297
TOF (–)	Threshold value for overland flow. Overland flow is only generated if the relative moisture content in the lower zone storage is larger than TOF	0	0.9	0.0000327
TIF (–)	Threshold value for interflow. Interflow is only generated if the relative moisture content in the lower zone storage is larger than TIF	0	0.9	0.86
TG (–)	Threshold value for recharge. Recharge to the groundwater storage is only generated if the relative moisture content in the lower zone storage is larger than TG	0	0.9	0.87
CK_{IF} (h)	Time constant for interflow from the surface storage. It is the dominant routing parameter of the interflow because $CK_{IF} >> CK_{1,2}$	500	1,000	560.3
$CK_{1,2}$ (h)	Time constant for overland flow and interflow routing. Overland flow and interflow are routed through two linear reservoirs in series with the same time constant $CK_{1,2}$	3	72	50
CK_{BF} (h)	Baseflow time constant. Baseflow from the groundwater storage is generated using a linear reservoir model with time constant CK_{BF}	500	5,000	3,999

Table 3.3 Evaluation of model performance for calibration and validation

Evaluation indexes	Calibration (2003–2007)	Validation (1998–2002)
Volume error (%)	−0.007	−11.304
Mean|Qsim-Qobs|/Qobs	2.00	1.31
R^2	0.811	0.826
EI	0.809	0.807

value of interest. The shape of a flow-duration curve in its upper and lower regions is particularly significant in evaluating the flow characteristics. The projected inflows show a very steep curve in the high-flow region which is expected for

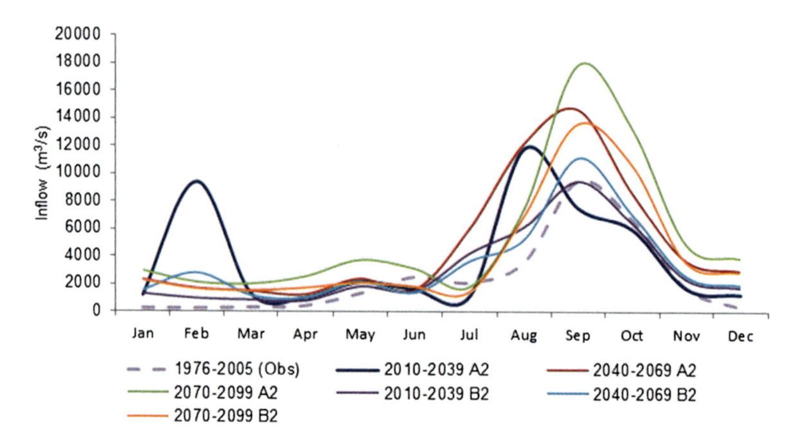

Fig. 3.10 Inflow to Ubolratana dam for different time windows under future climate

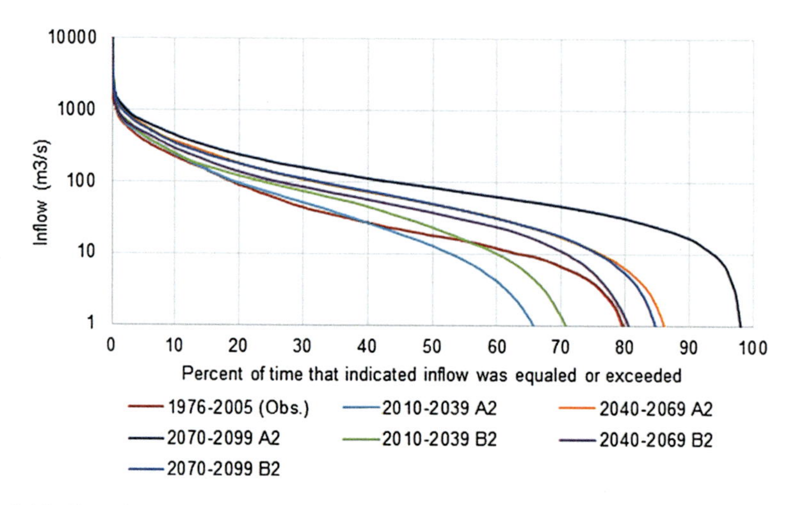

Fig. 3.11 Flow duration curve of daily inflow to Ubolratana dam

rain-caused floods on this basin (Fig. 3.11). In the low-flow region, the beginning of 21st century exhibit high percentage of no flow which is relatively higher than the historical time period, until the mid of the century where there are more low flows in each step. In addition, an inclined curve indicates that moderate flows are not sustained throughout the year due to natural inflow regulation, or because a small groundwater capacity cannot sustains the base flow to the stream.

3.4.4.3 RRV Analysis for Future Climate

Annual inflow data of Ubolratana dam from 1970 to 2008 is used as the level of water for baseline period. Low annual inflow or less than 20th percentile of this

Table 3.4 Results obtained by RRV analysis for the inflow to Ubolratana dam

	Observed	SRES A2			SRES B2		
	1976–2005	2010–2039	2040–2069	2070–2099	2010–2039	2040–2069	2070–2099
Q20 (years)	7	9	1	0	6	5	3
Q80 (years)	7	8	18	26	14	18	17
Reliability, C_R	0.533	0.433	0.367	0.133	0.333	0.233	0.333
Resiliency, C_{RE}	0.643	0.529	0.368	0.115	0.300	0.174	0.350
Vulnerability time, C_V (years)	3	3	7	11	6	12	8

period is assumed to cause drought whereas high inflow or more than 80th percentile may cause flood. Flow between 20th and 80th percentiles are assumed to be the appropriate inflow for which the dam operators safely. Table 3.4 illustrates the RRV analysis results obtained from the study based on the projected future runoff. Evidently number of low annual inflow (Q20) is expected to decrease in the future however; the number of high annual inflow (Q80) is projected to increase under both the emission scenarios. Surprisingly, although for certain years in the future is projected to have high intense rainfall, yet low annual inflows are observed in the corresponding year. This is probably due to higher rate of evapotranspiration attributing to the high temperature and the contribution of more percolation in the aquifers. In addition our analysis show higher and lower rainfall in the future will decrease resilience and reliability however increases the vulnerability for both A2 and B2 scenarios. The analysis also show that 2050s time period for B2 scenario is the most vulnerable contributing to vulnerability for 12 years. However, for A2 scenario, 2080s time window is more vulnerable relative to other time periods. Nonetheless, the future is ascertained to be more severe and the reservoir operation rule is necessary to be reviewed for future climate.

3.5 Conclusions

The present study analyses the future inflow and the Resilience, Reliability, Vulnerability (RRV) analysis of the flow to Ubolratana dam in Thailand under future climate for 2020s, 2050s and 2080s under A2 and B2 climate scenarios. Climate data were collected from 39 meteorological stations and streamflow data from 26 gauging stations in the upstream of the dam. Bias correction of the climate data was done for 42 grids from upstream and downstream of the dam for the RCM PRECIS. Power law transformation was applied to correct the maximum and minimum temperatures whereas, non-liner method of multi-day window for correction of coefficient of variation was used to correct the precipitation. Further the bias corrected temperature and precipitation was used as input for the hydrological model MIKE-11 NAM to simulate the future inflow. An additional RRV analysis of the simulated inflow is also done to analyze the vulnerability of the dam under future climate.

The results suggest an increase in the precipitation for both scenarios under future climate and all time windows considered. A significant increase of 36, 35 and 42 % in average annual rainfall is expected for 2020s, 2050s and 2080s under A2 scenario whereas, 25, 19 and 40 % for B2 scenarios for the corresponding time periods. Similarly, climate change is expected to induce higher temperature for the future climate with a magnitude of 0.50, 1.36 and 2.46 °C for 2020s, 2050s and 2080s under A2 scenario for mean maximum temperature and 0.51, 1.13 and 1.85 °C for the corresponding time windows under B2 scenario. Likewise mean annual minimum temperature is expected to increase by 0.61, 1.71 and 3.13 °C for 2020s, 2050s and 2080s respectively under A2 scenario and 0.60, 0.43 and 2.30 °C for B2 scenario for corresponding time periods relative to the baseline period of 1976–2005. The simulated runoff changes are driven by combined effects of rainfall changes and their seasonality. Simulated inflows shows increase for all period and both emission scenarios, with the greatest change occurring in period 2080s for A2 emission scenarios. Most of extreme changes are in low and moderate flow quantile ranges. Compared to the historical period, the number of high annual inflow will increase while the number of low annual inflow will decrease. The RRV criteria show that with the increasing rainfall in future will contribute to lower resilience and reliability whereas higher vulnerability. The results of this study show an increase in the volume of inflow for all the projected period which will affect the storage of dam. Therefore, the appropriate planning and management should be ready to counteract this problem for the future.

References

Artlert K, Chaleeraktrakoon C, Nguyen VTV (2013) Modeling and analysis of rainfall processes in the context of climate change for Mekong, Chi and Mun river basins (Thailand). J Hydro-environ Res 7(1):2–17

Chulalongkorn (2012) Thailand water account (2005–2007). Water Resources System Research Unit. Faculty of Engineering. Chulalongkorn University, Bangkok

DHI Water & Environment (2007) MIKE 11: A modelling system for rivers and channels, Reference Manual

Fowler HJ, Kilsby CG, O'Connell PE (2003) Modeling the impacts of climatic change and variability in the reliability, resilience and vulnerability of a water resource system. Water Resource Res. doi:10.1029/2002WR001778

Hunukumbura PB, Tachikawa Y (2012) River discharge projection under climate change in the Chao Phraya river basin, Thailand using MRI_GCM3.1S dataset. J Meteorol Soc Jap 90A:137–150

Intergovernmental Panel on Climate Change (2007) Climate change 2007: The physical science basis—Summary of policymakers. Cambridge University Press, Cambridge, UK and New York

Klemeš V (1986) Operational testing of hydrologic simulation models. Hydrolog Sci J 31:13–24

Lauri H, de Moel H, Ward PJ, Räsänen TA, Keskinen M, Kummu M (2012) Future changes in Mekong River hydrology: Impact of climate change and reservoir operation on discharge. Hydrol Earth Syst Sci 16:4603–4619

Leander R, Buishand TA (2007) Resampling of regional climate model output for the simulation of extreme river flows. J Hydrol 332:487–496

Manomaiphiboon K, Octaviani M, Torsri K, Towprayoon S (2013) Projected changes in means and extremes of temperature and precipitation over Thailand under three future emissions scenarios. Clim Res 58:97–115

Raje D, Mujumdar PP (2010) Reservoir performance under uncertainty in hydrologic impacts of climate change. Adv Water Res 33:312–326

Royal Irrigation Department, RID (2002) Study plan to support the development of primary water resources and improvement of irrigation project for 9th economic development planning. 25th River Basin Status Report, Bangkok, Thailand

Sharma D, Babel MS (2013) Application of downscaled precipitation for hydrological climate change impact assessment in the Ping River Basin of Thailand. Clim Dyn 41:2589–2602

Sharma D, Gupta AD, Babel MS (2007) Spatial disaggregation of bias corrected GCM precipitation for improved hydrologic simulation: Ping River Basin, Thailand. Hydrol Earth Syst Sci 11:1373–1390

Sivakumar B (2011) Global climate change and its impacts on water resources planning and management: Assessment and challenges. Stoch Environ Res Risk Assess 25:583–600

Terink W, Hurkmans RTWL, Torfs PJJF, Uijlenhoet R (2010) Evaluation of a bias correction method applied to downscaled precipitation and temperature reanalysis data for the Rhine basin. Hydrol Earth Syst Sci 14:687–703

Teutschbein C, Seibert J (2012) Bias correction of regional climate model simulation for hydrological climate-change impact studies: Review and evaluation of different methods. J Hydrol 456–457:12–29

Chapter 4
Assessment of Climate Change Impacts on Flood Hazard Potential in the Yang River Basin, Thailand

Abstract This study aims to analyze the impacts of climate change on flood hazard in Yang River Basin under future climatic scenarios with coupling of a physically-based distributed hydrological model, Block-wise application of TOPMODEL using Muskingum-Cunge flow routing (BTOPMC) and hydraulic model, HEC-RAS. The bias corrected outputs of a regional climate model PRECIS were used to construct climate scenarios for the 2020s, 2050s and 2080s. The extreme runoff pattern and synthetic inflow hydrographs for 25, 50 and 100 year return flood were derived from an extreme flood of 2007 which were then fed into HEC-RAS model to generate the landuse and flood inundation relationship for the basin. Results indicate that croplands are being mostly affected up to 188 km^2 for 100 year return period in case of baseline period. Furthermore, total area inundated for the corresponding return periods for baseline period are 205, 224 and 240 km^2. This amount of inundated area is projected to occur corresponding to 16 year flood in the period of 2020s under A2 scenario. Similarly the 25 year probable flood event is expected to have the most relative change (+30.90 %) for 2050s for same scenario and in case of B2 scenario, it is expected to be +30.97 % of the total inundated area for 2080s relative to baseline period. The probable increase in flood hazard under climate change scenarios threatens the increased inundation of croplands area and indicates the potential damage in food production and its impacts on livelihood of local people.

Keywords Climate change · Flood hazard · Thailand · Hydrological modelling · Hydraulic modeling

4.1 Introduction

Yang river basin is one of the most flood prone basins in Northeast Thailand (Kuntiyawichai et al. 2011). Several studies on climate change impact assessment and flood management strategies have been conducted on its main basin, Chi in

S. Shrestha, *Climate Change Impacts and Adaptation in Water Resources and Water Use Sectors*, Springer Water, DOI 10.1007/978-3-319-09746-6_4

recent years (Chaleeraktrakoon and Khwanket 2013; Artlert et al. 2013; Kuntiyawichai et al. 2011). These study reports that climate change is consistent and it has strong implications on the basin scale hydrological cycle. Other studies done globally indicate the altercated meteorological variables have great potential to change the frequency and intensity of extreme events specially floods (Dobler et al. 2012; Viviroli et al. 2011). The increase in temperature accelerates the evapotranspiration process which further influences the precipitation amount and ultimately contributes in modification of seasonal runoff. The present intra-annular variability in the amount of runoff is expected to shift under climate change in future at many regions of the world including Thailand (Dobler et al. 2010).

Aside from the projected changes in the hydrological regime, the climate change will also have implications on the extreme events. Studies have demonstrated that flood intensity is highly sensitive to temperature in many parts of the world (Prudhomme et al. 2013; Menzel et al. 2002; Panagoulia and Dimou 1997). Several other studies also have argued that climate has been a contributing factor to flood risk by raising the precipitation amount relative to the average annual rainfall (Fleming et al. 2012; Hirabayashi et al. 2008). Therefore basin scale assessment of climate change impacts on flood plays a key role in evaluation of adaptation and mitigation strategies for sustainable flood risk management.

Literature suggests that studies on assessing impact of climate change on extreme events have been less investigated and possess higher uncertainty (Dobler et al. 2012). In addition, whatsoever the research has been conducted, primary focus is on the basin of developed nations (Bauwens et al. 2011; Prudhomme et al. 2010; Steele-Dunne et al. 2008). Also focusing on Asian countries, many studies on floods induced by climate change has been conducted on several basins in China (Li et al. 2013; Zheng et al. 2012; Yang et al. 2012). This implies less focus on basins of developing countries lying on the tropical regions which are evidently more susceptible to floods where the region has already high precipitation and hydrologic cycle is highly interlinked and sensitive to its components (Kite 2001). Although considerable studies on floods have been conducted in northeast of Thailand yet merely a handful of studies has been done on climate influence on extreme events (Jothityangkoon et al. 2013; Hunukumbura and Tachikawa 2012). Despite of several flood events in Yang river basin most of the studies focus on the management practices and socio-economic impacts of floods (Keokhumcheng et al. 2012; Dutta 2011; Hungspreug et al. 2000). Hence the study of climate change impact on flood hazard is important at basin scale in Thailand.

Another important factor that has decisively influenced the climate change impact studies is the use of Regional Climate Model (RCM) dataset for the future climate projection without bias correction (Cloke et al. 2013). Although RCMs perform nested dynamic downscaling to the outputs of the General Circulation Models (GCMs), yet the spatial resolution makes the data unreliable for basin scale impact assessment studies and is necessary to be bias corrected (Muerth et al. 2013). A few studies have been conducted so far on analysis of different downscaling techniques with emphasis on extreme events. A comparison study of six downscaling technique with three RCMs suggests both statistical and dynamic downscaling tends to have similar bias. However, the choice of method of downscaling depends on variables to be downscaled (Schmidli et al. 2007). Leander and Buishand (2007) satisfactorily used the power law transformation method for RCM outputs at Western Europe for estimation of extreme events.

In summary, there is an immediate need of downscaling climate extremes in flood prone regions and analysis of climate change impacts on floods in Yang river basin, Thailand. Therefore the present study is conducted to analyze the impact of climate change on flood hazard in Yang River Basin with the following objectives: (i) to develop rainfall-runoff model to represent the Yang River Basin, (ii) to design synthetic hydrographs with return periods of 25, 50 and 100 years with regard to future climate conditions, and (iii) to simulate flood hazard potential representing with return periods of 25, 50 and 100 years with regards to future climate change scenarios. The output of this research can be used as a benchmark for climate resilient flood management plan for Yang river basin which has been severely damaged in recent past.

4.2 Materials and Methods

4.2.1 Study Area and Data Description

The Yang River basin has a drainage area of approximately 4,145 km^2 which receives an average annual rainfall of 1,390 mm (Fig. 4.1).The annual relative humidity and temperature are around 71 % and 26.7 °C in the basin, respectively. The basin is influenced by two prominent wind systems, the northeast and southwest monsoons which are responsible for the rainfall patterns and temperature variations. The northeast monsoons, the dry cold wind picks up some moisture from the northeast, take place from mid-October to early February. The southwest monsoons begin around mid-May and fade down by mid-September. In addition to monsoons, the Yang River basin also faces tropical storms. The tropical depressions mainly come from the South China Sea. Consequently, the high moisture travelling over the water surface causes the heavy rain during rainy season (Artlert et al. 2013).

Topographically, the basin is characterized by the Phu Phan mountain range at a relatively high elevation of around 600 masl, with the Yang River as the major river that flows through Kalasin province, and meets Chi River at Yasothon province. The land use in this basin consists of agriculture (70 %), forest (25 %), urban (2 %), water bodies (1.2 %) and others (1.8 %).

The DEM, a digital representation of ground surface topography, was constructed from geometric data acquired from Thailand Land Development Department and Royal Irrigation Department. The geometric data consists of point elevation, 2-m interval contours, stream elevation, road elevation and Yang River boundary. Those geometric data were used to generate DEM of 20 m resolution.

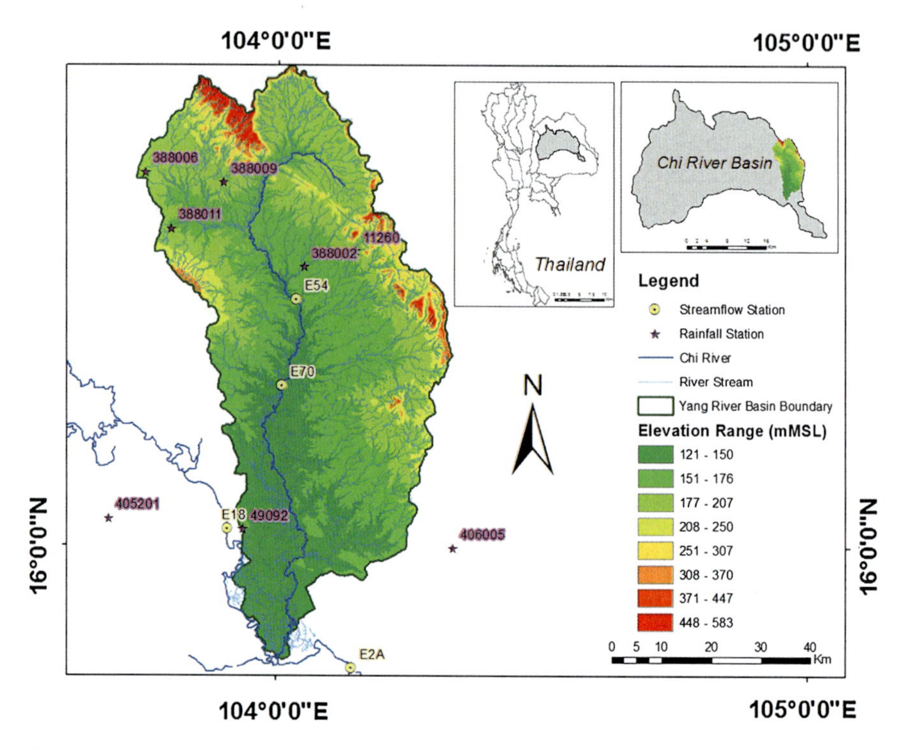

Fig. 4.1 Location map of study area with hydro-meteorological stations in Yang River Basin, Thailand

A land cover/land-use map for this study area was obtained from Thailand Land Development Department and classified based on the International Geosphere Bioshpere Programme (IGBP). Soil classification map for the basin is acquired by using the Food and Agriculture Organization (FAO) digital soil map of the world.

A three step modelling approach i.e. (i) correcting the biasness of the large-scale atmospheric data, (ii) hydrological and hydraulic modelling for flood inundation and flood hazard analysis and (iii) change analysis was adopted in this study. Figure 4.2 gives an overview of the methodology adopted in this study. Firstly, BTOPMC and HEC-RAS model were calibrated on the basis of observed climate variables; while in the second step, the output from the RCM were bias corrected. Thereafter, the future precipitation data was in turn used as a forcing to the hydrological and hydraulic model, which generated runoff series and flood inundation areas based on present and future climate conditions. In the end, the two runoff series and inundation areas were compared and a change analysis was carried out.

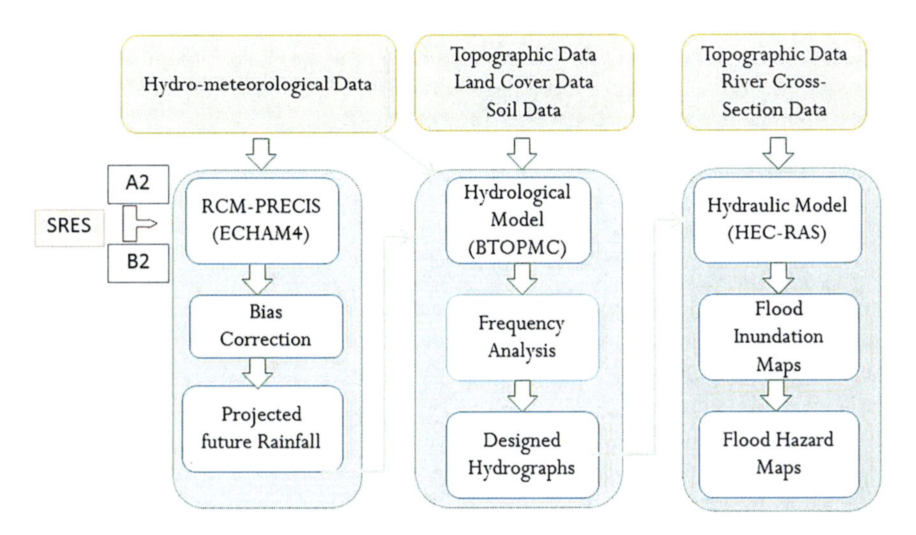

Fig. 4.2 Schematic representation of the methodology used in flood hazard map creation

4.2.2 SRES Scenarios and GCMs

The climate scenarios used in this investigation were derived from the regional climate model (RCM) entitled providing regional climates for impact studies (PRECIS). The simulations are based on the Special Report on Emission Scenarios (SRES), A2 and B2 scenarios. The A2 scenario is the highest emission scenario and B2 scenario generally represents the medium emission during the 21st century. The B2 scenario is 'environmental and social sustainability' world in future with the CO_2 concentration of 800 ppm and projected temperature of 1.4–3.8 °C. The A2 scenario is 'regionally oriented economic development' world with CO_2 concentration of 1,250 ppm and projected temperature of 2.0–5.4 °C (IPCC 2007).

PRECIS uses general circulation model (GCM) ECHAM4 as its lateral boundaries under the prescribed IPCC A2 and B2 scenarios. It is also capable to give an output of daily precipitation, temperature, solar radiation and wind speed which is produced by the "Southeast Asia System for Analysis, Research and Training" (SEA START) Regional Center. The spatial resolution of the downscaled data by PRECIS is 0.2° × 0.2° covering 2,225 grids for entire Mekong river basin. Further details on PRECIS can be derived from Jones et al. (2004). Due to the extensive application of PRECIS in climate change impact assessment studies in Southeast Asian countries since last 5 years (Mainuddin et al. 2013; Ty et al. 2012; Mainuddin et al. 2011; Tuan and Chinvanno 2011; Hoanh et al. 2010; Graiprab et al. 2010) and its ability to replicate the present observed climate in the region, the model output was selected for this study.

Climate scenarios information was transferred from RCM to the hydrological model and frequency analysis was performed on the simulated hydrological scenarios. The results of the hydrological modelling were used as input for flood inundation analysis.

4.2.3 Bias Correction of RCM Data

Bias correction was performed for each sub-basin, since the biases in temperature and precipitation fluctuate spatially. The goal of bias correction was to carry out daily time-series of modified precipitation and temperature at any point throughout the domain of interest.

Since temperature record at a station is approximately normally distributed, the correction was done by shifting and scaling the values to adjust the mean and variance as followed by Leander and Buishand (2007). Correction of precipitation was done by existing well accepted method of power law transformation in which the coefficient of variation (CV) and the mean of the data are corrected. Further details on temperature and precipitation correction used in this study can be can be retrieved from Leander and Buishand (2007).

4.2.4 Goodness-of-Fit Indicators of RCM Outputs

The criteria used for evaluating RCM outputs and observed climate variables were coefficient of determination, percent bias and index of agreement.

4.2.4.1 Coefficient of Determination (R^2)

This index measures the strength of a linear relationship between the two variables. The value of R^2 varies between 0 and 1 which indicates how much of the observed data matched by the simulated data.

4.2.4.2 Volume Bias (VB)

This statistical parameter measures the average propensity and mass balance of computed data comparing to the observed data. The optimal value of VB is zero. Positive values show model underestimation bias; meanwhile, negative values specify model overestimate bias. VB is expressed as in Eq. 4.1

$$VB = \left[\frac{\sum_{i=1}^{n} (X_{obs,i} - X_{sim,i})}{\sum_{i=1}^{n} X_{obs,i}} \right] \tag{4.1}$$

Where, X_{obs} and X_{sim} are the observed and simulated climate variables at time i, respectively, and n is the number of time step.

4.2.4.3 Index of Agreement (*d*)

This statistical index indicates the degree of model prediction error (Willmott 1981). The index of agreement is dimensionless which diverges between 0 and 1 (Eq. 4.2). A determined value of 1 shows a perfect fit between the observed and simulated values; 0 shows no agreement at all. The index of agreement describes the ratio between the mean square error and the "potential error" (Willmott 1984).

$$d = 1 - \frac{\sum_{i=1}^{n} (X_{obs,i} - X_{sim,i})^2}{\sum_{i=1}^{n} \left(\left| X_{sim,i} - \overline{X} \right| + \left| X_{obs,i} - \overline{X} \right| \right)^2} \tag{4.2}$$

Where, X_{obs} and X_{sim} are the observed and simulated climate variables at time i, respectively, and n is the number of time step.

4.2.5 Hydrological Modelling

The hydrological simulations were carried out using Block-wise use of TOPMODEL with Muskingum-Cunge method (BTOPMC) model. It is an extension of the TOPMODEL concepts (Beven et al. 1995), which is developed in order to overcome the limitations of using the TOPMODEL for large river catchments. For large river catchments, spatial heterogeneity and timing of flow to outlet are the important factors. For representing spatial variability in BTOPMC, a catchment is composed of grid cells, which can be divided into sub-catchments, where each sub-catchment is considered as a block or a unit. The effective precipitation for any sub basin, i, and soil layer, k, and time, t is calculated based on Eq. 4.3.

$$Re(i,t) = Ro(i,t) - Srmax(i) - Ep(i,t) \tag{4.3}$$

Where, Re represents the effective rainfall, Ro indicates total precipitation, $Srmax$ is the maximum storage capacity in the root zone and Ep denotes evapotranspiration. BTOPMC integrates Shuttleworth-Wallace model to calculate the

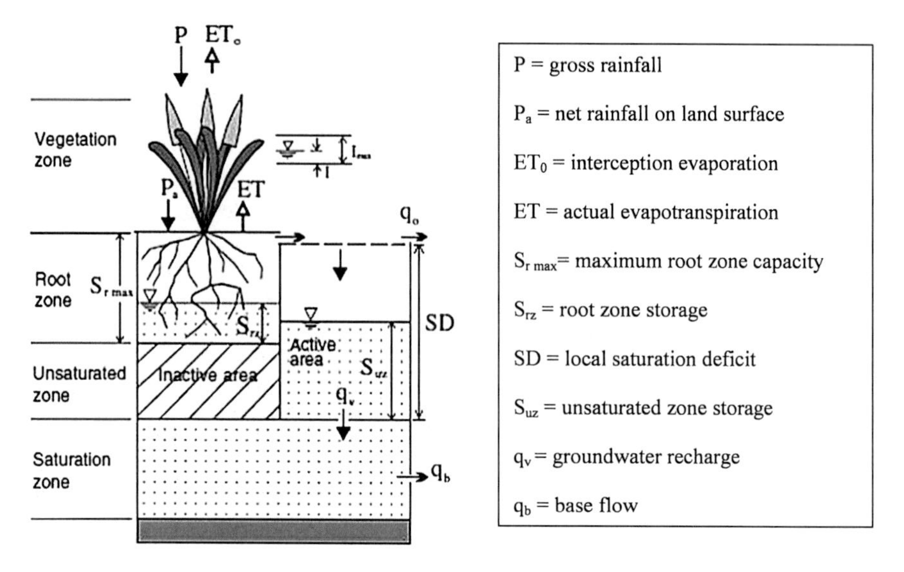

Fig. 4.3 Soil water balance and runoff generation in BTOPMC for each grid cell

potential evapotranspiration. Figure 4.3 gives the details of the runoff generation for each grid cells in BTOPMC model. Further details are available on Ao et al. (2006) and Ishidaira et al. (2005). BTOPMC model has been successfully applied in many basins worldwide with satisfactory performance (Manandhar et al. 2013; Phan et al. 2010; Bao et al. 2010; Wan et al. 2009; Shrestha et al. 2007). In addition, it has also been applied successfully to Mekong River basin which validates the model's ability to represent the hydrology of the basin (Kiem et al. 2008; Hapuarachchi et al. 2008).

The model was calibrated by adjusting saturated transitivity, decay coefficient and rooting depth for the period of 2002–2006 in order to yield maximum Nash-Sutcliffe efficiency criteria (Nash and Sutcliffe 1970) and R^2 whereas minimize volume error (Bao 2006). The calibrated parameters were used as input for the validation period of 1997–2001 in order to check for the best goodness-of-fit with the observed discharge at Kuchinarai, Kalasin station (E54).

4.2.6 Frequency Analysis and Designed Hydrographs

In order to analyze and design the frequency of probable maximum stream flow, annual maximum runoffs of the gauging station from 1980–2009 and the annual maximum runoffs for the future periods of 2020s, 2050s and 2080s were simulated using distributed hydrological model. The application of the runoff results are to illustrate the extreme flood frequency given by the extreme value (EV) I distribution, also called the Gumbel distribution method. For the flood inundation modeling under climate change conditions, several return periods of stream flow such as 25, 50 and 100 years at Yang River basin are considered.

Designed synthetic hydrographs were developed by applying two methods, the dimensionless hydrograph method and the flood frequency analysis. The dimensionless hydrograph was derived from the extreme runoff pattern in 2007. For flood frequency analysis, Gumbel's distribution was selected to analyze the historical runoff data recorded from 1980–2009. Moreover, the Gumbel probability distribution was also used to predict future runoff simulated by the distributed hydrological model by using the bias corrected future climate data. The frequency analysis provided the probable maximum runoff with 25, 50 and 100 year return period from the Gumbel probability distribution. The dimensionless hydrographs are used to derive synthetic inflow hydrographs for the above mentioned return periods.

4.2.7 Hydraulic Modelling

Flood inundation areas were simulated using HEC-RAS, a 1-D hydraulic model innovated by the US Army Corps of Engineers, under unsteady flow conditions (U.S. Corps of Engineers 2002). The topography of channel and floodplain information was derived from HEC-GeoRAS software which is used as an extension in ArcGIS, for processing geospatial data for use with HEC-RAS. It extracts topography data from an existing digital elevation model (DEM) as input files to HEC-RAS. The geospatial data include river, reach, cross sectional cut lines, cross sectional surface lines, cross sectional bank stations, downstream reach lengths for the left overbank, main channel, and right overbanks, and cross sectional roughness coefficients. The essential geometric data contains stream centerlines and stream cross section and these are prepared using the HEC-GeoRAS. It also allows the import of the prepared data into HEC-RAS model.

Discharge and water level values were set as upstream and downstream boundaries. The upstream boundary in this study is at Kuchinarai station (E.54)—a stream-flow gauging station. In order to take backwater effect into the Yang River model, the water level at the junction between Yang River and Chi River was also calculated by HEC-RAS model. Based on the availability of spatial extent of water and flood level in the basin A. Pon Thong (E.70) was selected to setup the model. The model calibration was done by adjusting Manning's roughness coefficient (n) in order to give best goodness-of-fit for the period from 1 July to 30 November 2005. The selected validation period was from 1 July to 30 November 2007.

4.2.8 Flood Frequency Analysis

The objective of flood frequency analysis was to relate the magnitude of floods to their frequency of occurrence using probability distribution. In order to commence the objective, first the calculation of the statistical parameters of the proposed distribution was done by the method of moments from the given data. In this study the annual maximum runoffs for four different time periods (1990s, 2020s, 2050s and 2080s) were calculated for different return period floods.

4.2.9 Flood Hazard Mapping

Flood hazard mapping is to determine areas with a probability of a flooding event for a defined return period (Han 2011). With the results of hydraulic calculations, the flood outline can be calculated. The main step is to calculate the inundation area by subtracting the digital terrain model from the water level based on the results produced by the 1-D hydraulic model.

The degree of flood hazard depends on several hydrological factors such as velocity and inundated depths. Since this study applied the 1-D hydraulic model, the hazard index was assigned with corresponding to different inundated depths. The degree of flood hazard was classified into four hazard categories based on inundation depth classes corresponding to three critical inundated depths 0.6, 1.0 and 3.5 m as suggested by Tu and Tingsanchali (2010) which is demonstrated in Table 4.1.

4.2.10 Change Analysis

The relationships between the magnitude and frequency of extreme events were derived from the daily discharge data. Therefore, the annual maximum stream flow data in this study were analyzed with the Gumbel Distribution (Extreme Value type I).

Table 4.1 Hazard index for depth of flooding

Depth of flooding (m)	Degree of flood hazard	Description	HI
D > 3.50	Very high	"Extreme danger: flood zone with deep fast flowing water"	4
1.00–3.5	High	"Danger: flood zone with deep fast flowing water"	3
0.60–1.00	Moderate	"Danger: flood zone with deep fast flowing water"	2
D < 0.60	Low	"Flood zone with shallow flowing water or deep standing water"	1

The alteration in the area of inundation corresponding to the floods of various return periods relative to the baseline period was analyzed to evaluate the implication of climate change on flood occurrence.

4.3 Results and Discussion

In this section, the performance of (i) downscaling of GCMs (ii) performance of BTOPMC simulation and (iii) hydrological analysis are presented along with the projected flood hazard inundation maps for future climate change scenarios for a different return period flood.

4.3.1 Model Performance Evaluation

4.3.1.1 Performance of Bias Correction of RCM Data

The RCM outputs forced by ECHAM5 were bias corrected by applying the power law transforms for rainfall data and the linear approach for temperature data. The observed data for the 30-year period of 1976–2005 were used as a baseline in this study due to available climate data.

Correcting Rainfall

The intra-seasonal and spatial precipitation pattern suggests low coefficient of variation (CV) for daily rainfall which can be attributed to the higher amount of rainfall in the basin. As a result, parameters *a* and *b* were determined by considering the CVs of multi-day rainfall. Table 4.2 summarizes the performance of the CVs of multi-day rainfall amounts matched those of the corresponding days from observed

Table 4.2 Goodness-of-fit indicators of observed rainfall and RCM outputs including the CVs of multi-day rainfall amounts

Station ID	Goodness-of-fit indicators			Multi-day
	R^2	VB	d	CVs
				Days
11260	0.99	0.01	0.99	25
388002	0.99	0.01	0.99	
388006	0.99	0.03	0.99	
388009	0.99	0.02	0.99	
406005	0.98	0.03	0.99	35
49092	0.99	0.02	0.99	
388011	0.96	0.08	0.98	45

Table 4.3 Performance of BTOPMC in Yang River basin

Period	*NSE* (%)	r^2	*Vr*
Calibration (2002–2006)	62.80	0.63	0.97
Validation (1997–2001)	66.45	0.68	1.12

rainfall. The goodness-of-fit indicators used to choose the amount of multi-days are coefficient of determination (R^2), volume bias (*VB*) and index of agreement (*d*).

The bias correction method employed to adjust the uncorrected RCM involves correcting for the mean, standard deviation and the coefficient of variation. Since the rainfall was found to vary spatially and temporally, the parameters a and b determined by considering the moving windows of 25, 35, 45, 55 and 65-day were different in each rainfall station.

The analysis of multi-day rainfall amounts for 25, 35, 45, 55 and 65-days suggests the corrected RCM for mean; standard deviation and CV of 25-day rainfall amounts (CV25) have the best goodness-of-fit indicators at four stations. In addition, it is also observed that the lesser number of days considered shows better representation of the observed amount of rainfall for instance 65-day rainfall indicates highest variability followed by 55 days in the rainfall intensity and regime. Furthermore, it was also noted that the uncorrected RCM indicates wetter period for June and July relative to the observed rainfall which is contradictory.

4.3.1.2 Performance of the BTOPMC

During the calibration, BTOPMC uses observed rainfall data of 2002–2006 as input into a systematic search for model parameters which produce the best goodness-of-fit between the simulated discharge and observed discharge at Kuchinarai, Kalasin (E54). Thereafter, 5 years from 1997 to 2001 were used for validation purposes where both calibration and validation periods cover extreme flood events in the study area. The performance of the model during calibration and validation period is summarized in Table 4.3 and Fig. 4.4. The low flow periods are underestimated during calibration whereas overestimated during validation for the first 3 years followed by underestimation. In addition, it can also be noticed that the model is able to simulate the low peaks very well however it is unable to fetch the high peak flow during heavy rainfall events during validation period. Nonetheless, the model evaluation by NSE, R^2 and Vr indicates that the model performs reasonably well in the basin.

4.3.1.3 Performance of Hydrological Analysis and Design

Annual maximum runoffs of 1980–2009 and for the future period of 2020s, 2050s and 2080s simulated from the distributed hydrological model were used to design frequencies of the annual runoffs. The implication of the runoff results are to characterize the flood frequency given by the extreme value (EV) I distribution.

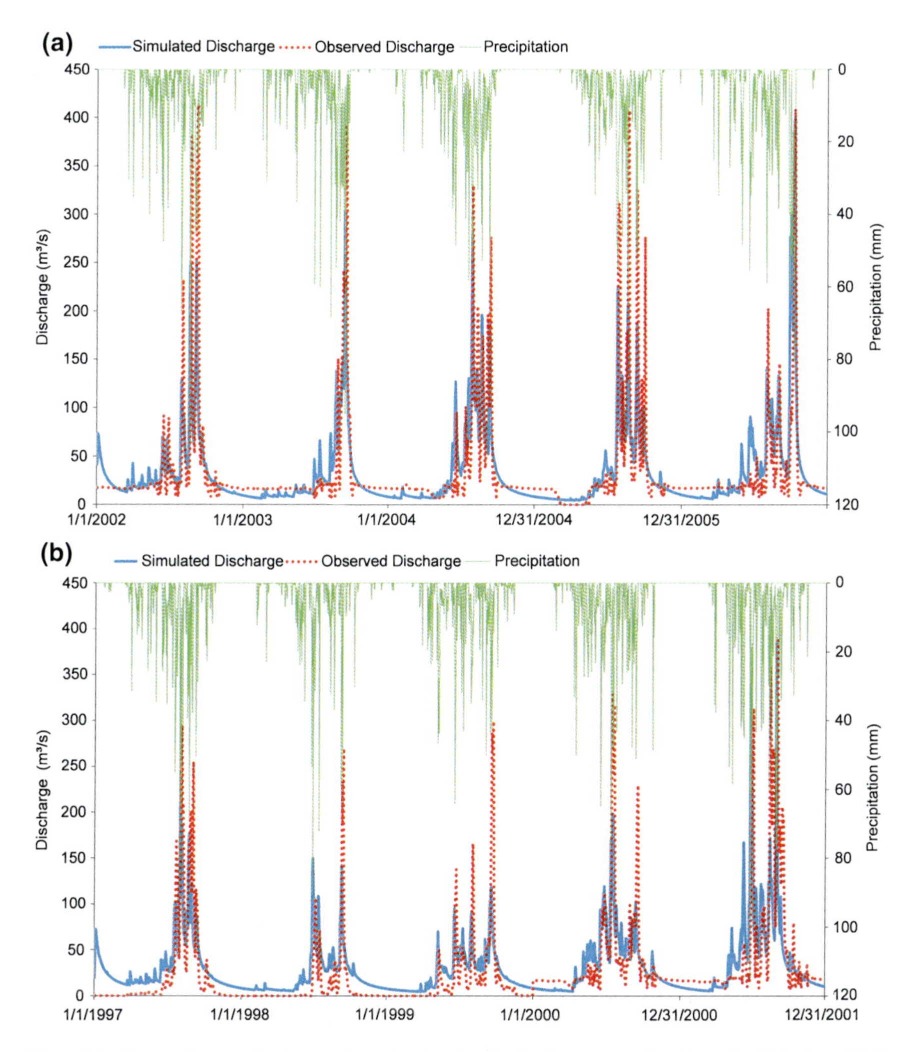

Fig. 4.4 Comparison of observed and simulated discharge at Kuchinarai, Kalasin (E54) **a** calibration period (1 Jan 2002 to 31 Dec 2006) and **b** validation period (1 Jan 1997 to 31 Dec 2001)

Flood Frequency Analysis

Table 4.4 illustrates the probable peak discharge at Kuchinarai station under A2 and B2 scenarios. Maximum change in discharge is observed for 100 years return period for all the three future time periods under both scenarios. Furthermore, 2050s time window shows highest change in magnitude of discharge for A2 scenario which can be attributed to the future change in precipitation. This analysis indicates 2050s are expected to have higher intensity flood with all the return period considered (discussed later). In addition analysis for B2 scenario explains the highest

Table 4.4 Probable peak discharge for A2 and B2 scenarios at Kuchinarai station

Time period	Discharge in m³/sec for different return period (T)						
	Scenario	25 years	Relative change (%)	50 years	Relative change (%)	100 years	Relative change (%)
Baseline		552.33		611.63		670.49	
2020s	A2	701.96	27.09	809.07	32.28	915.40	36.53
	B2	905.02	63.85	1,058.35	73.04	1,210.55	80.55
2050s	A2	1,053.07	90.66	1,207.72	97.46	1,361.23	103.02
	B2	730.37	32.23	831.11	35.88	931.10	38.87
2080s	A2	954.23	72.76	1,078.22	76.29	1,201.30	79.17
	B2	917.91	66.19	1,045.52	70.94	1,172.20	74.83

relative change of discharge for 50 and 100 year return period is observed for 2020s compared to the baseline period. Moreover, a considerable change in the flood for intra-return period is also noticeable for 2020s and 2080s. Although the change is not significant relative to each other however it is dynamic compared to the baseline discharge for all return periods.

Figure 4.5a suggests under A2 scenario, floods of higher intensity are expected to be more frequent. In addition, floods with lower return periods and intensity are expected to be less for the 2020s and 2050s relative to the baseline period. Moreover, floods of higher return period are expected to have increase in magnitude for 2050s. The projection of rainfall in the basin by RCM can be attributed to this expected increase in magnitude of flood for 2050s. Based on the result of flood frequency analysis under B2 scenario, the probable flood events in the future period of the 2020s and 2080s are more extreme relative baseline period and 2050s (Fig. 4.5b). The intra-return period variability for maximum discharge in case of 2020s and 2080s are minimal for extreme floods. In addition, the lower intensity floods with short return periods are expected to be less compared to the base line period. The change in precipitation projected by RCM under climate change is the contributing factor for this pattern in discharge.

Figure 4.6a shows the relative change in probable peak discharge between baseline and the projected discharges under A2 scenario. The probable peak discharges of 25-year return period is observed to increase by 27, 91 and 73 % in the future periods of 2020s, 2050s and 2080s respectively. For 50 year return period, the probable peak discharges increases by 32, 97 and 76 % in the future periods of 2020s, 2050s and 2080s respectively. For 100 year return period, the probable peak discharges increase by 37, 103 and 79 % in the corresponding future periods.

Figure 4.6b shows the relative change in probable peak discharge between baseline and the projected discharges under B2 scenario. The probable peak discharges of 25-year return period is projected to increase by 64, 32 and 66 % in the future periods of 2020s, 2050s and 2080s respectively relative to baseline period. Similarly for 50-year return period, the probable peak discharges increase by 73, 36

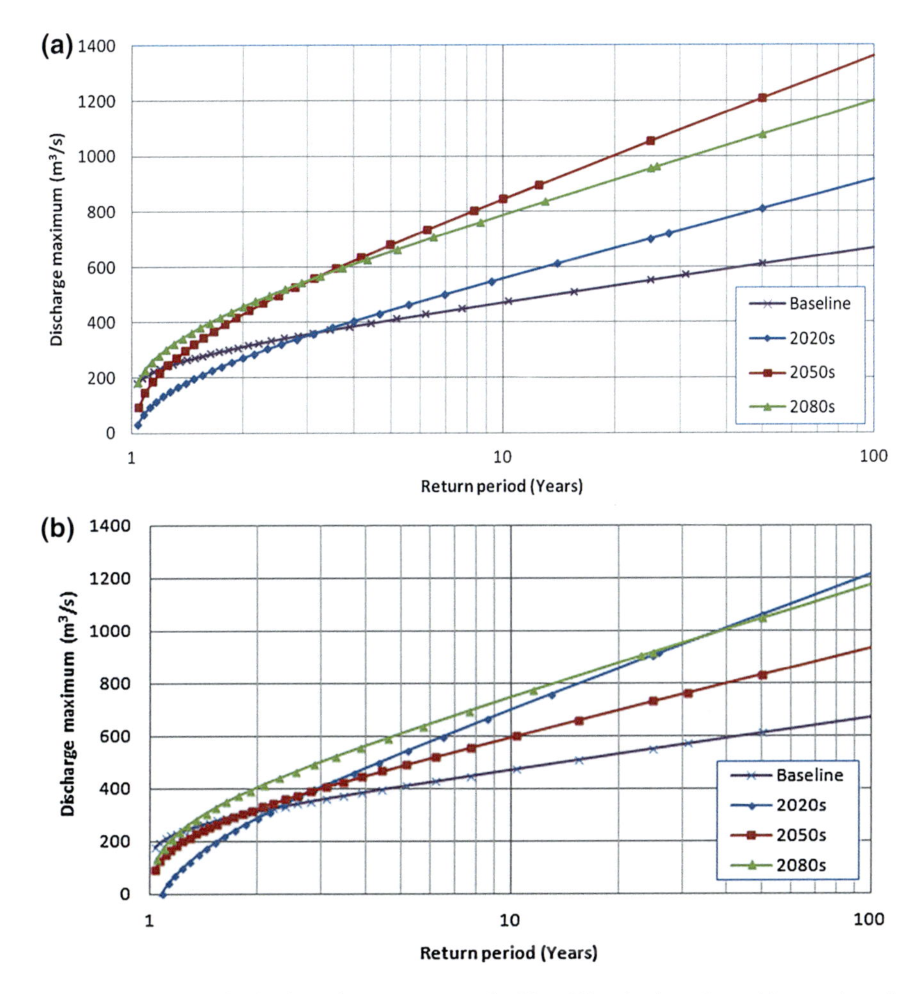

Fig. 4.5 Frequency distribution of extreme events for Yang River basin under **a** A2 scenario and **b** B2 scenario

and 71 % in the same periods respectively. For 100-year return period, the probable peak discharges increase by 81, 39 and 75 % in the look ahead periods of 2020s, 2050s and 2080s respectively.

Designed Hydrographs

Synthetic hydrograph generated for different return periods at Kuchinarai station for baseline period suggests maximum flood is expected for 100 year return period followed by 50 and 25 year. In addition, it is also observed that the time to peak for all the return period floods follows same trend as that of the observed hydrograph. Moreover, maximum variation in the magnitude of the designed hydrographs is observed at the peak.

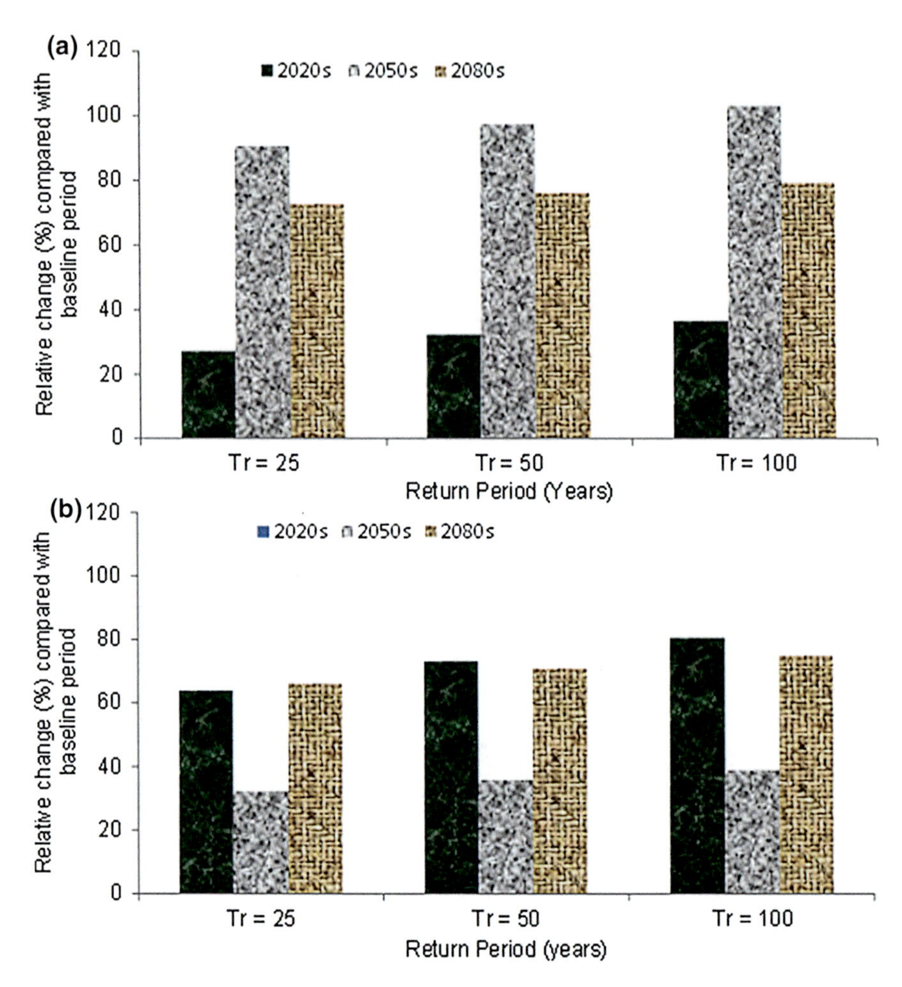

Fig. 4.6 Relative changes in probable peak discharge compared to baseline period **a** A2 scenario and **b** B2 scenario

4.3.1.4 Performance of HEC-RAS Model

Model Evaluation Based on Statistical Indices

The calibration and validation statistics suggests the model simulates the observed flood in good agreement (Table 4.5). Figure 4.7 demonstrates the simulated water level compared to the observed water level at Pon Thing (E.70). The validation results suggest the maximum observed water surface level at 140.61 masl on October 7, 2007. The simulated water surface level is 139.29 masl on the same day illustrating the model's ability to fetch the time to peak at the same time.

Table 4.5 Performance of HEC-RAS model in Yang River

Period	r^2	*RPE*	*VB*
Calibration (2005)	0.96	0.71	−0.01
Validation (2007)	0.92	0.94	−0.01

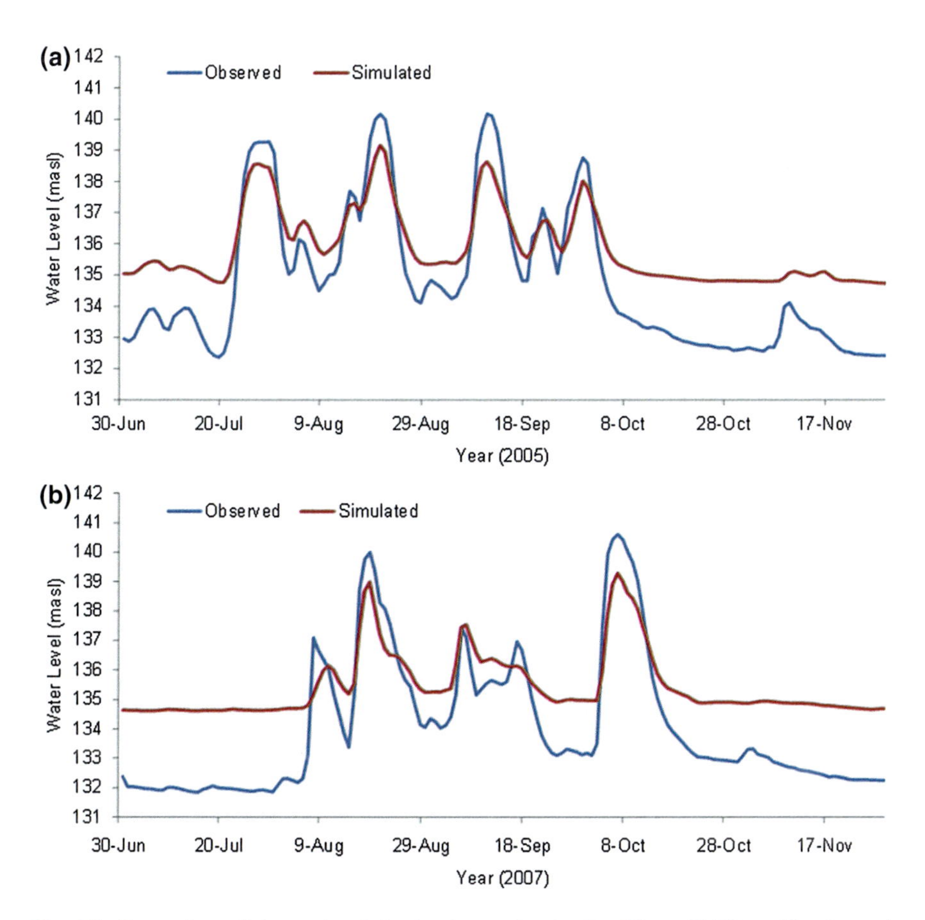

Fig. 4.7 Comparison of observed and simulated water level at Pon Thong (E.70) **a** calibration and **b** validation

4.3.2 Projected Changes in Floods

4.3.2.1 Flood Inundated Areas for Baseline Period (1980–2009)

Table 4.6 illustrates the areas of inundation under different degrees of flood hazard. Based on the assigned degree of flood hazard, although all class of flood hazards shows an increase, yet the spatial coverage of high floods is observed to be

Table 4.6 Flood inundation areas in different return period floods for baseline period

Degree of flood hazard	Simulated flood area in km^2 (in %)		
	25-year	50-year	100-year
Low	25.44 (12.38)	28.01 (12.51)	28.79 (12.02)
Moderate	19.25 (9.37)	17.32 (7.73)	17.73 (7.40)
High	141.66 (68.95)	151.63 (67.71)	154.31 (64.42)
Very high	19.10 (9.30)	26.99 (12.05)	38.69 (16.15)
Total	205.44	223.94	239.51

maximum under present climate. In case, of moderate and high flood events highest spatial coverage is observed for 25 year return period. In addition, very high flood is noted to have maximum coverage of 38.69 km^2 which includes 16.15 % of total area for 100-year return period. It is also noteworthy that the trend in the area of inundation for degree of flood hazard reduces from 25- to 100-years return period for low, moderate and high floods. However, for very high flood, area of inundation is observed to follow increasing trend for the corresponding return periods. For the total area of inundation, flood of 100-year return period has the maximum area of inundation with 239.51 km^2 implying the severity of extreme floods in the region under present climate.

Further analysis on land use of the inundated area for the extreme floods suggests significant increase from 159.46 to 188.47 km^2 in the area of croplands in case of 25–100 year return period (Table 4.7). It is also evident that, only change in inundation area for high and very high floods influences the total area for croplands indicating agricultural vulnerability. This implies that farming lands acts as retention areas for flood water in the region. It can also be noticed that forests experiences high and very high floods of all return periods compared to low and moderate. In addition, the spatial extent of high floods declines with increased return period and tends to increase for very high.

4.3.2.2 Changes in Future Flood Inundation Area

The simulation was carried out for the present and future scenarios of extreme rainfall events. Future flood inundation areas were simulated for 2020s, 2050s and 2080s with respect to A2 and B2 scenarios (Table 4.8). It can be noticed in case of A2 scenario, the spatial coverage of flooded area increases noticeably for all return periods and three time windows considered relative to the baseline period. Interestingly, total area of flood inundation follows a declining trend from 284 to 268 km^2 and 303 to 268 km^2 for 50 and 100 year return periods respectively in case of 2080s which can be attributed to the change in projected rainfall. Extreme flood events are observed to have maximum area of inundation (95 to 189 km^2) for 2050s for all return periods. Moreover, lower intensity floods are observed to have a declination in the area of inundation under future climate for A2 scenario.

Table 4.7 Flood inundation depth generated by HEC-RAS—land use relationship for baseline period in 25, 50 and 100 year flood

Return period	Land use	Low D < 0.6 (km^2)	Moderate 0.6 < D < 1.0 (km^2)	High 1.0 < D < 3.5 (km^2)	Very high D > 3.5 (km^2)	Total area (km^2)
25-year	Forest	0.88	0.56	9.81	4.06	15.30
	Croplands	21.69	16.52	114.47	6.78	159.46
	Grasslands	1.67	1.45	16.04	8.23	27.38
	Urban	1.19	0.72	1.35	0.03	3.30
	Water bodies	1.64	2.21	38.96	11.48	54.29
	Total	25.44*	19.25*	141.66*	19.10*	205.44*
50-year	Forest	1.05	0.64	7.85	6.55	16.08
	Croplands	24.28	14.74	125.86	10.68	175.56
	Grasslands	1.41	1.16	15.92	9.72	28.21
	Urban	1.27	0.78	2.00	0.03	4.08
	Water bodies	0.97	1.28	35.17	17.41	54.84
	Total	28.01*	17.32*	151.63*	26.99*	223.94*
100-year	Forest	1.26	0.63	6.66	8.30	16.84
	Croplands	24.47	15.17	130.20	18.63	188.47
	Grasslands	1.27	1.02	14.87	11.71	28.88
	Urban	1.79	0.90	2.58	0.05	5.32
	Water bodies	0.73	0.77	29.89	23.84	55.23
	Total	28.79*	17.73*	154.31*	38.69*	239.51*

* water bodies ignored.

Furthermore, high magnitude flood of 100 year return period is expected to decline for 2050s and 2080s relative to baseline period for corresponding scenario.

Simulation for B2 scenario suggests, although the total area of inundated increases significantly from 251–284 km^2 relative to 205–239 km^2 for baseline period under future climate, yet lower coverage of area under flood is observed for 2050s. Large-scale flood can be observed to peak at 2050s with inundation area ranging from 142 to 152 km^2 for the three return periods. Analysis of output for 2050s illustrates a considerable increase in low, medium and high flood however; very high flood shows a significant decline relative to 2020s and 2080s. Intra-scenario analysis also suggests the severity of flood is less for the B2 scenario relative to that of A2 for the corresponding period.

Further studies done in Mekong delta region of Vietnam suggest under climate change high and very high floods are expected to be more intense (Dinh et al. 2012). Moreover, studies on streamflow under climate change on Mekong river basin suggests the frequency of peak discharge is expected to change from 75 year return period to 25 years for future climate by 2045s (Lauri et al. 2012) which implies the results of the present research are in line with other studies done in the region.

Table 4.8 Flood hazard areas estimated from various return periods under A2 and B2 scenarios in Yang River basin

Time period	Return period	Scenarios	Low* D < 0.6 (km^2)	Moderate* 0.6 < D < 1.0 (km^2)	High* 1.0 < D < 3.5 (km^2)	Very high* D > 3.5 (km^2)	Total area* (km^2)
2020s	25	A2	27.64	18.93	154.19	45.86	246.62
		B2	16.06	17.53	141.77	93.02	268.38
	50	A2	22.25	18.45	147.16	73.56	261.41
		B2	13.50	11.09	131.04	122.33	277.97
	100	A2	17.05	17.68	142.68	89.92	267.33
		B2	13.68	9.48	116.75	144.44	284.34
2050s	25	A2	15.41	17.05	140.96	95.51	268.93
		B2	26.61	19.70	152.31	53.19	251.81
	50	A2	13.67	9.48	116.71	144.51	284.37
		B2	21.13	18.16	146.03	77.62	262.94
	100	A2	12.26	9.08	92.67	189.62	303.63
		B2	16.60	17.61	142.29	91.30	267.79
2080s	25	A2	16.06	17.53	141.77	93.02	268.38
		B2	15.35	17.01	140.89	95.82	269.07
	50	A2	15.50	17.18	141.02	94.98	268.68
		B2	13.56	11.34	131.83	120.89	277.62
	100	A2	13.54	11.16	130.93	121.54	277.17
		B2	13.66	9.57	118.03	142.56	283.80

*Water bodies ignored

4.4 Conclusions

Floods are natural phenomenon not uncommon in Yang River basin occurring annually. The present study is divided into three parts, projection of future peak discharge, flood inundation and land use relationships for the past severe floods and projection of future flood with different return periods. Future climate data derived from PRECIS RCM were further downscaled to stations by power law transformation followed by simulation of hydrology of basin by using physically-based distributed hydrological model BTOPMC. The flood hazard potential was projected by using the 1-D hydraulic model HEC-RAS.

Based on the results obtained from simulations by hydrological model, no particular trend was obtained for 25, 50 and 100 year return period discharge in case of both A2 and B2 scenarios. However, the relative change varies from +27 to +103 % and +32 to +80 % for A2 and B2 scenarios respectively indicating lower intra-scenario variability for the latter. Simulation of floods for the present climate suggests the croplands are affected maximum with high and very high floods covering an area of 159, 175 and 188 km^2 for 25, 50 and 100 year return period respectively which implies agricultural sector is under threat in the region. Area inundated by flood under future climate suggests 100 year return period floods are most severe. In addition, for a particular return period of flood, a shift from very high under baseline period to moderate flood is expected for future climate under A2 scenario. Furthermore, 100 year return period flood is expected to be 25 years in near future (2020s) for both climate scenarios which signifies the severe threat of flood in future under climate change. The output of this study not only indicates the severity of flood in the region but also focus on the land use affected under present climate which implies flooding of cultivation land indicates potential damage in food production and negative effects on the livelihoods of local people. So, proper land use and risk-based design of hydraulic structures must be integral part of mitigation plan when addressing vulnerabilities to reduce future flood damages in the basin.

Acknowledgements The authors would like to appreciate Dr. Jun Magome and the BTOPMC Development Team of Yamanashi University, Japan for providing the model to conduct this study.

References

Ao T, Ishidaira H, Takeuchi K, Kiem AS, Yoshitari J, Fukami K, Magome J (2006) Relating BTOPMC model parameters to physical features of MOPEX basins. J Hydrol 320:84–102

Artlert K, Chaleeraktrakoon C, Nguyen VTV (2013) Modeling and analysis of rainfall processes in the context of climate change for Mekong, Chi and Mun River Basins (Thailand). J Hyd Env Res 7(1):2–17

Bao HJ, Wang LL, Li ZJ, Zhao LN, Zhang GP (2010) Hydrological daily rainfall-runoff simulation with BTOPMC model and comparison with Xin'anjiang model. Water Sci Eng 3 (2):121–131

Bao WM (2006) Hydrologic forecasting, 3rd edn. China Water Conservancy and Water Power Press, Beijing (in Chinese)

Bauwens A, Sohier C, Degré A (2011) Hydrological response to climate change in the Lesse and the Vesdre catchments: contribution of a physically based model (Wallonia, Belgium). Hydrol Earth Syst Sc 15(6):1745–1756

Beven KJ, Lamb R, Quinn PF, Romanowicz R, Freer J (1995) Topmodel. In: Singh VP (ed) Computer models of watershed hydrology, Water Resources Publications, pp 627–668

Chaleeraktrakoon C, Khwanket U (2013) A statistical downscaling model for extreme daily rainfalls at a single site. World Environmental and water resources congress 2013: showcasing the future—Proceedings of the 2013 congress, pp 1058–1067

Cloke HL, Wetterhall F, He Y, Freer JE, Pappenberger F (2013) Modelling climate impact on floods with ensemble climate projections. Quarterly J Royal Met Soc 139:282–297

Dinh Q, Balica S, Popescu I, Jonoski A (2012) Climate change impact on flood hazard, vulnerability and risk of the Lonng Xuyen Quadrangle in the Mekong Delta. Intl J River Basin Management 10(1):103–120

Dobler C, Bürger G, Stötter J (2012) Assessment of climate change impacts on flood hazard potential in the Alpine Lech watershed. J Hydrol 460–461:29–39

Dobler C, Stötter J, Schöberl F (2010) Assessment of climate change impacts on the hydrology of the Lech Valley in northern Alps. J Water Clim Change 1(3):207–218

Dutta D (2011) An integrated tool for assessment of flood vulnerability of coastal cities to sea-level rise and potential socio-economic impacts: a case study in Bangkok, Thailand. Hydrol Sci J 56 (5):805–823

Fleming KM, Tregoning P, Kuhn M, Purcell A, McQueen H (2012) The effect of melting land-based ice masses on sea-level around the Australian coastline. Aust J Earth Sci 59(4):457–467

Graiprab P, Pongput K, Tangtham N, Gassman PW (2010) Hydrologic evaluation and effect of climate change on the Samat watershed, Northeastern Region, Thailand. Intl Agric Eng J 19 (2):12–22

Han D (2011) Flood risk assessment and management. Bentham

Hapuarachchi HAP, Zhou MC, Kiem AS, Geogievsky MV, Magome J, Ishidaira H (2008) Investigation of the Mekong River basin hydrology for 1980–2000 using YHyM. Hydrol Process 22:1246–1256

Hirabayashi Y, Kanae S, Emori S, Oki T, Kimoto M (2008) Global projections of changing risks of floods and droughts in a changing climate. Hydrol Sci J 53(4):754–772

Hoanh CT, Jirayoot K, Lacombe G, Srinetr V (2010) Comparison of climate change impacts and development effects on future Mekong flow regime. In: Swayne DA, Yang W, Voinov AA, Rizzoli A, Filatova T (eds) Fifth biennial meeting in modeling for environment's sake. Ottawa, Canada

Hungspreug S, Khao-uppatum W, Thanopanuwat S (2000) Flood management in Chao Phraya river basin. In: The Chao Phraya Delta: historical development, dynamics and challenges of Thailand's rice bowl: Proceedings of the international conference, Kasetsart University, Bangkok, Thailand, 12–15 December 2000

Hunukumbura PB, Tachikawa Y (2012) River discharge projection under climate change in the Chao Phraya River Basin, Thailand, using the MRI-GCM3. 1S Dataset. J Meteorol Soc Japan 90(A):137–150

IPCC (2007) Climate change 2007: the physical science basis. In: Solomon S, Qin D, Manning M, Marquis M, Averyt K, Tignor M, Miller HL, Chen Z (eds) Contribution of working group I to the fourth assessment report of the intergovernmental panel on climate change, Cambridge University Press, Cambridge and New York, NY

Ishidaira H, Takeuchi K, Magome J, Hapuarachchi P, Zhou M (2005) Application of distributed hydrological model YHyM to large river basins in Southeast Asia. In: Proceedings of International Symposium on Southeast Asian Water Environment, vol 3

Jones RG, Noguer M, Hassell DC, Hudson D, Wilson SS, Jenkins GJ, Mitchell JFB (2004) Generating high resolution climate change scenarios using PRECIS. Met Office Hadley Centre, Exeter

Jothityangkoon C, Hirunteeyakul C, Boonrawad K, Sivapalan M (2013) Assessing the impact of climate and land use changes on extreme floods in a large catchment. J Hydrol 490:88–105

Keokhumcheng Y, Tingsanchali T, Clemente RS (2012) Flood risk assessment in the region surrounding the Bangkok Suvarnabhumi Airport. Water Intl 37(3):201–217

Kiem AS, Ishidaira H, Hapuarachchi HP, Zhou MC, Hirabayashi Y, Takeuchi K (2008) Future hydroclimatology of the Mekong River basin simulated using the high-resolution Japan Meteorological Agency (JMA) AGCM. Hydrol Process 22(9):1382–1394

Kite G (2001) Modeling the Mekong: hydrological simulation for environmental impact studies. J Hydrol 253(1–4):1–13

Kuntiyawichai K, Schultz B, Uhlenbrook S, Suryadi FX, Griensven AV (2011a) Comparison of flood management option for the Yang River Basin, Thailand. Irrig Drain 60:526–543

Kuntiyawichai K, Schultz B, Uhlenbrook S, Suryadi FX, Corzo GA (2011b) Comprehensive flood mitigation and management in the Hi River Basin, Thailand. Lowland Tech Intl 13(1):10–18

Lauri H, de Moel H, Ward PJ, Räsänen TA, Keskinen M, Kummu M (2012) Future changes in Mekong River hydrology: impact of climate change and reservoir operation on discharge. Hydrol Earth Syst Sci 16:4603–4619

Leander R, Buishand TA (2007) Resampling of regional climate model output for the simulation of extreme river flows. J Hydrol 332:487–496

Li Y, Guo Y, Yu G (2013) An analysis of extreme flood events during the past 400 years at Taihu Lake, China. J Hydrol 500:217–225

Mainuddin M, Kirby M, Hoanh CT (2013) Impact of climate change on rainfed rice and options for adaptation in the lower Mekong Basin. Nat Hazards 66:905–938

Mainuddin M, Kirby M, Hoanh CT (2011) Adaptation to climate change for food security in the lower Mekong Basin. Food Sec 3:433–450

Manandhar S, Pandey VP, Ishidaira H, Kazama F (2013) Perturbation study of climate change impacts in a snow-fed river basin. Hydrol Process 27(24):3461–3474

Menzel L, Niehoff D, Bürger G, Bronstert A (2002) Climate change impacts on river flooding: a modelling study of three meso-scale catchments. In: Beniston M (ed) Climatic change: implications for the hydrological cycle and for water management, 10 Springer, Netherlands, pp 249–269

Muerth MJ, St-Denis BG, Ricard S, Velázquez JA, Schmid J, Minville M, Caya D, Chaumount D, Ludwig R, Turcotte R (2013) On the need for bias correction in regional climate scenarios to assess climate change impacts on river runoff. Hydrol Earth Syst Sci 17:1189–1204

Nash JE, Sutcliffe JV (1970) River flow forecasting through conceptual models 1: discussion of principles. J Hydrol 10:282–290

Panagoulia D, Dimou G (1997) Sensitivity of flood events to global climate change. J Hydrol 191 (1–4):208–222

Phan TTH, Sunada K, Oishi S, Sakamoto Y (2010) River discharge in the Kone River Basin (Central Vietnam) under climate change by applying the BTOPMC distributes hydrological model. J Water Clim Change 1(4):269–279

Prudhomme C, Crooks S, Kay AL, Reynard N (2013) Climate change and river flooding: part 1 classifying the sensitivity of British catchments. Clim Change 119:933–948

Prudhomme C, Wilby RL, Crooks S, Kay AL, Reynard NS (2010) Scenario-neutral approach to climate change impact studies: application to flood risk. J Hydrol 390(3–4):198–209

Schmidli J, Goodess CM, Frei C, Haylock MR, Hundecha Y, Ribalaygua J, Schmith T (2007) Statistical and dynamical downscaling of precipitation: an evaluation and comparison of scenarios for the European Alps. J Geophys Res. doi:10.1029/2005JD007026

Shrestha S, Mastola S, Babel MS, Dulal KN, Magome J, Hapuarachchi HAP, Kazama F, Ishidaira H, Takeuchi K (2007) The assessment of spatial and temporal transferability of a physically based distributed hydrological model parameters in different physiographic regions of Nepal. J Hydrol 347(1–2):153–172

Steele-Dunne S, Lynch P, McGrath R, Semmler T, Wang S, Hanafin J, Nolan P (2008) The impacts of climate change on hydrology in Ireland. J Hydrol 356(1–2):28–45

Tu VUT, Tingsanchali T (2010) Flood hazard and risk assessment of Hoang Long River basin, Vietnam. In: Proceedings of the International MIKE by DHI Conference, 2010
Tuan LA, Chinvanno S (2011) Climate change in the Mekong River Delta and key concerns on future climate threats. In: Stewart MA, Coclanis PA (eds) Environmental change and agricultural sustainability in the Mekong Delta, Adv Global Change Res 45:207–217
Ty TV, Sunada K, Ichikawa Y, Oishi S (2012) Scenario-based impact assessment of land use/ cover and climate changes on water resources and demand: a case study in the Srepok River Basin, Vietnam-Cambodia. Water Resour Manag 26:1387–1407
U.S. Corps of Engineers (2002) HEC-RAS river analysis system, Hydraulic Reference Manual. Hydraulic Engineering Center Report CPD-69, Davis, CA
Viviroli D, Archer DR, Buytaert W, Fowler HJ, Greenwood GB, Hamlet AF, Huang Y, Koboltschnig G, Litaor MI, Lopez-Moreno JI, Lorentz S, Schädler B, Schreier H, Schwaiger K, Vuille M, Woods R (2011) Climate change and mountain water resources: overview and recommendations for research, management and policy. Hydrol Earth Syst Sci 15(2):471–504
Wan YA, Yin F, Liu ZZ, Cui WC, Ao TQ (2009) Study on the effect of DEM spatial resolution and sampling algorithms on runoff simulation by BTOPMC. IAHS-AISH Publication, Washington, DC, 335:130–136
Willmott CJ (1984) On the evaluation of model performance in physical geography. In: Gaile GL, Willmott CJ (eds) Spatial statistics and models. D. Reidel, Dordrecht, Holland, pp 443–460
Willmott CJ (1981) On the validation of models. Phys Geogr 2:184–194
Yang C, Yu Z, Hao Z, Zhang J, Zhu J (2012) Impact of climate change on flood and drought events in Huaihe River Basin, China. Hydrol Res 43(1–2):14–22
Zheng P, Li Z, Bai Z, Wan L, Li X (2012) Influence of climate change to drought and flood. Disaster Adv 5(4):1331–1334

Chapter 5
Assessment of Climate Change Impacts on Irrigation Water Requirement and Rice Yield for Ngamoeyeik Irrigation Project in Myanmar

Abstract This study analyzes the temporal impacts of climate change on irrigation water requirement (IWR) and yield for rainfed rice and irrigated paddy respectively at Ngamoeyeik Irrigation Project (NIP) in Myanmar. Climate projections from two General Circulation Models (GCMs) namely ECHAM5 (scenario A2 and A1B) and HadCM3 (scenarios A2 and B2) were derived for NIP for future time windows (2020s, 2050s and 2080s). The climate variables were downscaled to basin level by using Statistical DownScaling Model (SDSM). Calibrated and validated AquaCrop v4.0 model was used to forecast the rainfed (May–October) yield and irrigation water requirement for irrigated paddy (November–April) under future climate. The analysis shows a decreasing trend in maximum temperature (−0.8 to +0.1 °C) for the three scenarios and three time windows considered; however, an increasing trend is observed for minimum temperature (+0.2 to +0.4 °C) for all cases. The analysis on precipitation also suggests that rainfall in wet season is expected to vary largely from −29 % (2080s; A1B) to +21.9 % (2080s; B2) relative to the average rainfall of the baseline period. A higher variation is observed for the rainfall in dry season ranging from −42 % for 2080s, B2; and +96 % in case of 2020s, A2 scenario. A decreasing trend of irrigation water requirement is observed for irrigated paddy in the study area under the three scenarios indicating that small irrigation schemes are suitable to meet the requirements. An increasing trend in the yield of rainfed paddy was estimated under climate change demonstrating the increased food security in the region.

Keywords Climate change · Crop modeling · Irrigation water requirements · Rainfed paddy yield · Myanmar

5.1 Introduction

Agriculture is the lifeline of Myanmar's economy which provides employment to 67 % of the working population and contributes to 58 % of Myanmar's gross domestic product (GDP) (UNDP 2011). Agricultural production is entirely

S. Shrestha, *Climate Change Impacts and Adaptation in Water Resources and Water Use Sectors*, Springer Water, DOI 10.1007/978-3-319-09746-6_5

dependent on the amount of water available in the field and hence, the dryland farmers of Myanmar mainly rely on rainfed farming for their livelihoods and the over dependence on rainfall makes Myanmar economy more vulnerable to climate change (ADB 2009). Rice is the major agricultural crop grown in Myanmar which covers 39.82 % of the total agricultural land area. In order to compensate the insufficient rainwater for irrigation; irrigated rice is being practiced in Myanmar in the new millennia as it provides assurance to the farmers for the summer rice production. A recent increase of 20.3 % in the irrigated areas has been observed within a span of 5 years (FAO 2009).

Climate change has become a global threat which has high potential to affect the water and agriculture sectors significantly (IPCC 2007; Molua 2009). In Southeast Asia, temperature is expected to rise by 1.87–3.92 °C and precipitation is anticipated to increase by 1–12 % by the end of the century as compared to the current condition (IPCC 2007). With the increased temperature and fluctuating precipitation, it is expected that water availability and crop productivity will decrease significantly in the future (Kang et al. 2009). It is also suggested that, climate change will have direct and indirect impacts on irrigation water requirement and yield of crops respectively (IFPRI 2009). The change in the yield is mostly expected due to the shift in the growth phase, photosynthesis capacity and increasing in the respiration and an increase in the water requirements. Moreover, various extreme climate events (e.g. floods, cyclones and heat waves) will have additional risk to the agricultural production (Alcamo et al. 2007).

Various modelling studies have confirmed that, with the increase of CO_2 concentration and temperature, a significant alteration in the productivity of rice has occured (Babel et al. 2011; Krishnan et al. 2007; Horie 2005; Erda et al. 2005; Inthavon et al. 2004). Also studies by Shrestha et al. (2013), Maeda et al. (2011) and De Silva et al. (2007) showed a remarkable increase in the irrigation water requirement (IWR) in the future as compared to the present climate. Some climate change impact assessment studies in agriculture also have found a significant increase in yield of many crops worldwide (Long et al. 2006; Fuhrer 2003; Amthor 1998). The increasing atmospheric CO_2 concentration influencing the temperature acts favorable for the spikelet formation in paddy and reduces the crop duration (Olesen et al. 2000). Agricultural crops in the tropical region are more sensitive to warming since they operate already close to the optimum temperature; however, in many regions a mild increase in warming with sufficient precipitation may have a net positive impact (Esterling and Apps 2005). Although various studies have been done in the Lower Mekong River Basin (LMRB) countries, very few studies has been conducted in Myanmar on climate change impact assessment.

The present study quantifies the change in the irrigation water requirement and rice productivity in an irrigation project area in southern region of Myanmar under different climate change scenarios. The outputs of this research will be highly useful for the farmers, policy makers and reservoir operators of the region to respond the adverse impacts of climate change on water resources and agriculture.

5.2 Study Area

The study area lies in the southern region of Myanmar with a catchment area of 414.5 km^2. Ngamoeyeik irrigation project is the largest project in the lower Myanmar and is located within latitude of 16°50′–17°30′N and longitude of 96°00′–96°30′E (Fig. 5.1). A substantial hilly region lies in the northern part of the basin with slope ranging from 4.5 to 9.0 % and a flat region exists in the south with average slope 0.3 %. The reservoir has an active storage capacity of 207 million cubic meters (MCM) and surface area of 44.5 km^2 irrigating 28,330 ha area. The irrigation scheme was designed to store the rainwater during monsoon and to utilize the storage water in the pre-monsoon season for double cropping. The reservoir has a potential capacity to irrigate of 28,330 ha area. It also acts as a reserve for the domestic water use during the dry periods for Yangoon city.

Sub-tropical climate dominates the region with an average annual rainfall of 2,700 mm and the monsoon season lies between May and October making the period suitable for rainfed paddy. The temperature ranges between 22.2 and 32.6 °C with extremes between 17.1 and 36.8 °C averaged from 1961 to 1990. The soil type

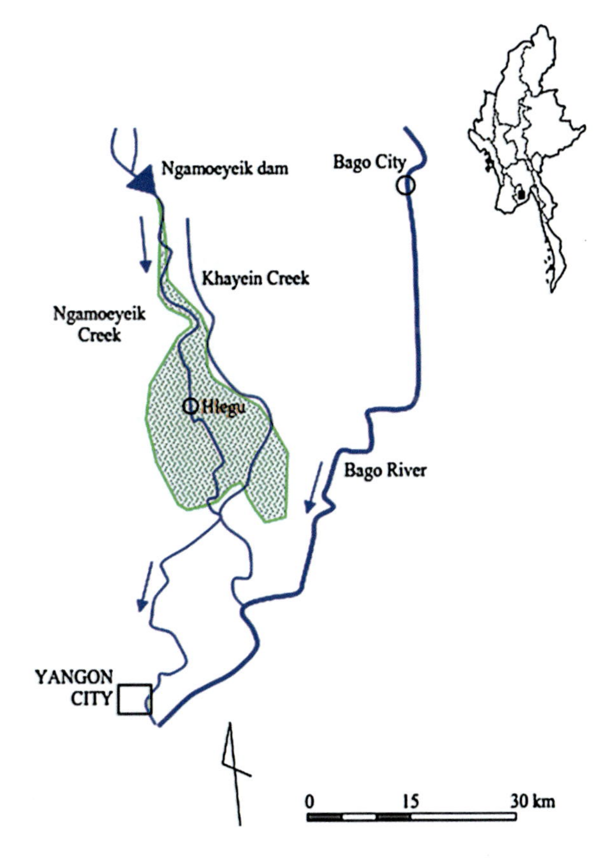

Fig. 5.1 Location map of Ngamoeyeik irrigation project in Myanmar

Table 5.1 Depth wise physical and chemical properties of soil at Ngamoeyeik irrigation project, Myanmar

Description	Soil depth (cm)		
	(0–10)	(10–20)	(20–30)
Clay (%)	33.00	67.00	8.66
Silt (%)	29.50	27.00	7.94
Sand (%)	33.00	58.50	7.98
Soil texture	Silty clay loam	Clay loam	Silty clay loam
FC (%)	44	39	42
PWP (%)	23	23	23
K (mm/day)	13	4	2.5
EC (μmoh)	67.00	27.00	58.50
pH	8.66	7.94	7.98
Specific gravity	2.67	2.67	2.67

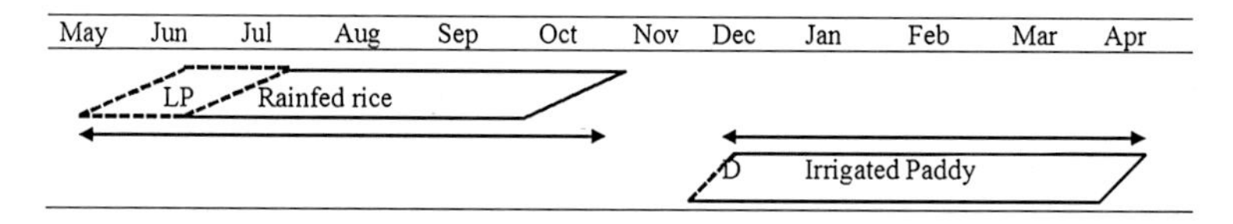

Fig. 5.2 Cropping calendar of rainfed rice and irrigated paddy showing post transplantation at Ngamoeyeik irrigation project in Myanmar

changes according to the depth. Table 5.1 elaborates the depth wise soil physical and chemical characteristics of soil in root zone.

In the study area, rice cultivation is generally practiced at a cropping intensity of 200 % with rainfed and irrigated conditions. Field preparation for the rainfed rice starts from middle of May and the crops are harvested by early November. In case of irrigated rice, transplantation is done by middle November and the paddy is harvested by middle of April. Figure 5.2 shows the cropping calendar followed for cultivating rice. Rice has five different phases namely, nursery, initial, development, reproductive and ripening stage (FAO 1998).

5.3 Materials and Methods

Two GCMs (ECHAM5 and HadCM3) were used to derive three SRES scenarios (A2, A1B and B2) of the future climate variables. They were downscaled at basin scale by statistical downscaling tool (SDSM) using the daily observed precipitation and temperature dataset obtained from the local meteorological station (Kabar Aye; Yangoon). The results obtained were used as input to the crop model, AquaCrop

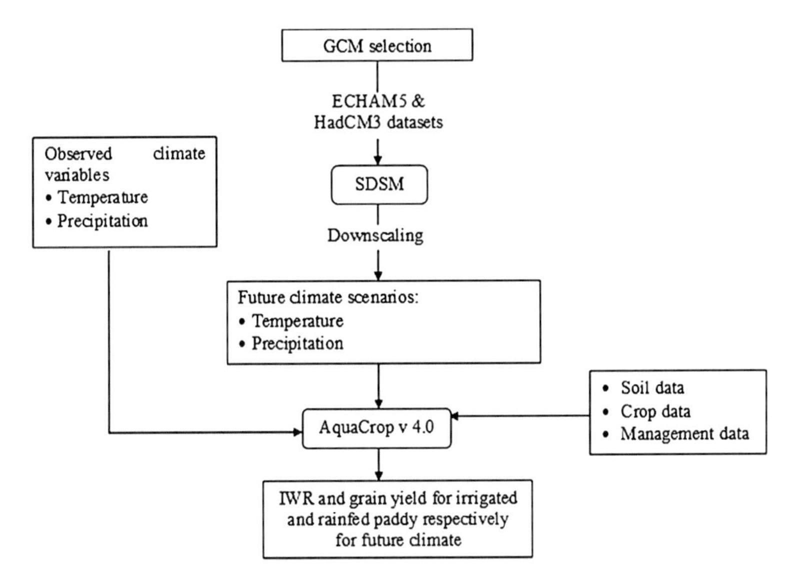

Fig. 5.3 The methodological framework for the study at Ngamoeyeik irrigation project in Myanmar

v 4.0 (Raes et al. 2009a), in order to estimate the rainfed rice yield and IWR for the irrigated paddy. Figure 5.3 demonstrates the methodological framework applied for this study. Similar approach was also used by Shrestha et al. (2013) and Maeda et al. (2011) to forecast the potential impacts of climate change on IWRs in Nepal and Kenya respectively.

5.3.1 Field Layout and Measurements

Field experiments for rainfed rice were conducted for the year 2010 and 2011 laid in randomized complete block design (RCBD) with the following treatments: no irrigation, W1; 25 % of irrigation water requirement (IWR), W2; 50 % of IWR, W3; and irrigation water at 100 % IWR, W4. Nitrogen was applied at two different rates N1: recommended dose of Nitrogen (50 kg/ha); N2: 60 % of recommended dose. Irrigation was provided by flooding method where water was derived from a reservoir and supplied by channels. The beds were surrounded by bunds such that they can store the irrigated water applied. Manawthukha which is a high yielding variety (HYV) was used for the field trials. Details of other mostly grown varieties are given in Table 5.2. Table 5.3 shows the water requirement based on crop growth stages for Manawthukha variety. The nurseries were transplanted at a depth of 2 cm and a spacing of 20 × 20 cm was provided for each plant (Young et al. 1998).

Table 5.2 Characteristics of rice varieties grown in Ngamoeyeik irrigation project, Myanmar

Name of rice varieties	Days to maturity	Plant height (cm)	Effective tillers/ plant	Length of panicle (cm)	Grains per panicle	1,000 grains weight (gm)	Yield (t/ha)
Manawthukha (HYV)	130–135	91–107	8–10	22	178	18.00	4–5
Yatanartoe (HYV)	105–120	76–92	7–9	24	133	28.18	4–5
Ayeyarmin (HYV)	140–145	137–152	10–12	25.4	160	24.00	3–4

Table 5.3 Stages of rice growth with crop water requirement at Ngamoeyeik irrigation project, Myanmar

Stage	Number of days	Water requirement (mm)[a]	% of total water requirement	Features
Nursery	50	60	4	From land preparation to transplanting
Initial	15	250	17	From transplanting to approx. 10 % of ground cover; dependent on crop variety
Development	45	550	38	From 10 % ground cover to effective full cover; initiation of flowering
Reproductive	30	450	31	From effective full cover to the start of maturity
Ripening	30	150	10	From start of maturity to harvesting
Total	170	1,460		

FAO (1998)

[a] http://agropedia.iitk.ac.in/?q=content/irrigation-water-management-paddy (accessed July, 2013)

The quantity of irrigation water applied in the field trials was calculated based on the IWR of paddy. IWR was calculated based on crop water use as in Eq. 5.1. Calculation of ETcrop is shown in later section however, effective rainfall (Pe) is calculated as given in Eq. 5.2. The volume of water was conveyed by the canals to the trial plots. The N fertilizer was applied in three split doses with one-third given at basal, second at 21 days after transplanting (DAT) and third at 42 DAT. The yield was measured at the physiological maturity of the paddy by selecting three middle rows of each experimental plot.

$$\text{IWR} = \text{ETcrop} - \text{Pe} \quad (5.1)$$

$$\begin{aligned} Pe &= 0.8P - 25 \quad \textit{if } P > 75\,\text{mm/month} \\ Pe &= 0.6P - 10 \quad \textit{if } P < 75\,\text{mm/month} \end{aligned} \quad (5.2)$$

where IWR, ETcrop, *Pe and P* are irrigation water requirement, evapotranspiration, effective rainfall, and rainfall, respectively, in mm.

5.3.2 Climate Change Scenario Generation

Forecasting the future climate variables is extremely difficult due to unpredictable greenhouse gas (GHG) accumulation in the atmosphere because of several natural and anthropogenic causes. Therefore in order to assess the potential impacts of future climate change on agriculture and water resources, we use various scenarios which we assume resembles the plausible future (IPCC 2007). They stipulate easy and correct way for determining the various driving forces leading to climate change and study the impact assessment studies along with uncertainty analysis.

According to the fourth assessment report by Intergovernmental Panel on Climate Change (IPCC), four emission scenarios have been considered namely A1, A2, B1 and B2 which are generated based on the different socio-economic and environmental consideration. A1 is further divided into A1FI, A1T and A1B which assumes the future will be more fossil intensive, non-fossil energy usage and balanced across all sources respectively (IPCC 2007). We have considered the worst case (A2), most optimistic (B2) and balanced usage of energy resources (A1B) scenario in our study since these three are the most presumptive scenarios in the context of Myanmar.

There are many General Circulation Models (GCMs) which projects different scenarios of the future climate based on the underlying assumed driving forces. Yet the availability of data, reliability on outputs and their resolution makes it important to do some analysis prior to use the GCM output for studies. It is unrealistic to rely on the output of a GCM with coarse resolution representing the future climate of any location with a good agreement (Yano et al. 2007). The four criteria based on which a GCM is selected for a particular region are vintage, resolution, validity and representativeness of the results (Smith and Hulme 1998).

The required scenarios i.e. A2 and A1B were extracted from ECHAM5/MPI-OM (European Centre-Hamburg model version 5/Max Planck Institute Ocean Model) whereas A2 and B2 were retrieved from HadCM3 (Hadley Centre Coupled Model version 3). Table 5.4 shows the detailed description of the two GCMs. These two GCMs have been used extensively in impact assessment studies in Southeast Asia

Table 5.4 Description of the GCMs used in this study

Model	Vintage	Country	Agency	Resolution (km × km)	
				Atmosphere	Ocean
HadCM3	1997	England	Hadley Center for Climate Prediction and Research/Met Office	96	73
ECHAM5/ MPI-OM	2005	Germany	Max Planck Institute of Meteorology	192	96

(Artlert et al. 2013; Nuorteva et al. 2010; Babel et al. 2011). Based on the outputs and conclusions of the above mentioned studies done and the robustness of their performance these two GCMs were selected for the study.

5.3.3 Transferring Coarse Resolution Climate Data to Basin Level

GCMs predict the climate variables at a global level which is not suitable for basin scale studies and moreover the regional features of local level are not amalgamated in GCMs (Russo et al. 1997). Downscaling is the process of transforming the GCM outputs to local level (IPCC 2007). Although, there are several methods of downscaling the coarse resolution data of GCMs to basin level viz., dynamical method, weather typing, stochastic weather generators and regression, the statistical downscaling method is preferred due to its cost effectiveness and its easiness to perform rapid assessments of localized climate (Bardosy and Plate 1992). Statistical DownScaling Model (SDSM) has become more accepted in recent years due to its applicability in wide region and simplicity of establishing relationship between predictor and predictand variables for future time zone (Wilby et al. 2002). Hence, SDSM package of decision support tool is used for this study to downscale maximum, minimum temperature and precipitation for the study area for (2010–2039) 2020s, (2040–2069) 2050s and (2070–2099) 2080s. Prior to forecasting the future climate variables, SDSM was calibrated based on observed data of 1961–1990 and then validated for the period of 1991–2000.

5.3.4 AquaCrop 4.0

AquaCrop is a windows based programme designed to simulate biomass and yield responses of field crops to different degrees of water availability under various soil conditions and climate change scenarios. It incorporates the soil-crop-atmosphere components through soil and water balance, atmosphere (rainfall, temperature, evapotranspiration and carbon dioxide concentration), crop characteristics (canopy cover, root depth, biomass production and yield) and field management practices (irrigation, fertility and agronomic practices) components (Raes et al. 2009a; Steduto et al. 2009). It also calculates the daily water balance and separates evapotranspiration into evaporation and transpiration. The model calculates the above ground biomass based on the Eq. 5.3 where it is a function of normalized water productivity, transpiration (factor of canopy cover), reference evapotranspiration (RE) and air temperature stress coefficient. The cumulative above ground biomass calculated is then converted into yield based on Eq. 5.4.

$$B = Ks_b \times WP^* \times \sum \frac{T_r}{ET_o} \tag{5.3}$$

$$Y = f_{HI} \times HI_O \times B \tag{5.4}$$

where B is above ground biomass in kg m^{-2}, Ks_b is air temperature stress coefficient, WP^* is normalized water productivity in kg m^{-2} mm^{-1} which is normalized for CO_2, type of product synthesized and soil fertility which is suggested to be kept 15 and 20 kg m^{-2} mm^{-1} for rice (Raes et al. 2009b), T_r is transpiration in mm and ET_o refers to evapotranspiration in mm. Y is referred as yield in kg m^{-1}, f_{HI} is adjustment factor for all the stress that affects the yield of crop, HI_O is the reference harvest index and B is the above ground biomass mentioned earlier in kg m^{-2}.

IWR is calculated based on the daily water flux computed by the water balance sub-component in the model. RE being computed by ET_o calculator was used as an input to the sub-component to calculate IWR. AquaCrop being a simple and less data intensive model was selected for this study because of its robust performance in various regions of the world (Abedinpour et al. 2012; Mkhabela and Bullock 2012; Geerts et al. 2009). Moreover, detailed crop phenology associated data availability being an issue for the study site. Although the model has several default values for crop parameters including rice, handful of parameters however, need to be tuned based on the local conditions, cultivars and management practices.

Crop water requirement is defined as the amount of water needed to compensate the evapotranspiration needs of a crop and is calculated as the difference between crop evapotranspiration and effective precipitation (Eq. 5.5). Crop evapotranspiration is a product of crop coefficient and RE (Eq. 5.6). However, RE is demonstrated as the rate of evapotranspiration by a hypothetical reference grass with crop height of 0.12 m, a fixed surface resistance of 70 s m^{-1} and albedo of 0.23. The standard method of calculation of RE is by FAO Penman-Monteith equation. More details can be found in Allen et al. (1998). In AquaCrop, effective precipitation is calculated considering the daily water balance, rainfall and monthly crop evapotranspiration as in Eq. 5.7 (USDA 1970).

$$CWR = ET_C - P_e \tag{5.5}$$

$$ET_C = K_c \times ET_O \tag{5.6}$$

$$Pe_m = \left(0.70917P_m^{0.82416} - 0.11556\right) \times 10^{0.02426ETc_m} \quad (5.7)$$

where CWR is crop water requirement in mm, ETc is crop evapotranspiration in mm, P_e is effective precipitation in mm, K_c is crop coefficient (unitless), ET_o is reference evapotranspiration in mm, and Pe_m, P_m, Etc_m are monthly effective rainfall, monthly rainfall and monthly crop evapotranspiration in inches (mm??) respectively.

The parameterization of crop model is evaluated based on Coefficient of Determination (R^2) which measures the strength of linear relationship between modeled and observed variables, standard deviation (σ^2) shows the variability of modeled and observed data compared to mean, Root Mean Square Error (RMSE) which measures the differences between values predicted by a model and observed values and Coefficient of Residual Mass (CRM) measures the tendency of the model over and underestimation.

5.4 Results and Discussion

5.4.1 Simulated and Observed Climate Data

The calibration process of SDSM involves developing relationship among meso-scale screened predictor variables and observed station data based on principle of multiple regressions. The summary of selected predictor variables and corresponding predictands for the study area meteorological station is given in Table 5.5. Meteorological data from 1961 to 1980 of Kabar Aye Station (Yangoon) (12 masl) located within 15 kms from experimental site is used for calibration of SDSM. It can be noticed that greater local variables influence the HadCM3 predictands compared to ECHAM5. Moreover, it can also be noted that there are different group of predictors that control the predictands of the two different GCMs. For instance, zonal velocity at 500 hPa (5_u), airflow strength at 850 hPa (8_f), relative humidity at 500 hpa (r500), relative humidity at 850 hpa (r850), near surface relative humidity (rhum) and surface specific humidity (shum) influences the maximum and minimum temperature of HadCM3. Whereas, 850 hpa geopotential height (P850), relative humidity at 500 hpa (r500) and relative humidity at 850 hpa (r850) influences the maximum and minimum temperature for ECHAM5. Similarly for precipitation, mean sea Level Pressure (mslp), 850 hpa geopotential height (P850), relative humidity at 850 hpa (r850) and 850 hPa divergence (8zh) are the drivers for HadCM3 while additional relative humidity at 500 hpa (r500), 850 hpa geopotential height (P850) with common predictor 850 hpa geopotential height (P850) influences ECHAM5.

Table 5.6 presents the performance of SDSM during the calibration and validation processes. During the calibration process, good comparison of the observed data with the data produced based on the generated transfer function, is obtained.

Table 5.5 Summary of selected predictor variables and their corresponding predictands of the GCMs at Ngamoeyeik irrigation project, Myanmar

GCM	Predictand	Predictors
HadCM3	Maximum temperature	Zonal velocity at 500 hPa (5_u)
		Airflow strength at 850 hPa (8_f)
		Relative humidity at 500 hpa (r500)
		Relative humidity at 850 hpa (r850)
	Minimum temperature	Near surface relative humidity (rhum)
		Surface specific humidity (shum)
		Zonal velocity at 500 hPa (5_u)
		Airflow strength at 850 hPa (8_f)
		Relative humidity at 500 hpa (r500)
		Relative humidity at 850 hpa (r850)
	Precipitation	Near surface relative humidity (rhum)
		Surface specific humidity (shum)
		Mean sea level pressure (mslp)
		850 hpa geopotential height (P850)
		Relative humidity at 850 hpa (r850)
		850 hPa divergence (8zh)
ECHAM5	Maximum temperature	850 hpa geopotential height (P850)
		Relative humidity at 500 hpa (r500)
		Relative humidity at 850 hpa (r850)
	Minimum temperature	850 hpa geopotential height (P850)
		Relative humidity at 500 hpa (r500)
		Relative humidity at 850 hpa (r850)
	Precipitation	850 hpa geopotential height (P850)
		Relative humidity at 500 hpa (r500)
		850 hpa geopotential height (P850)

Table 5.6 Performance of SDSM during calibration and validation at Ngamoeyeik irrigation project, Myanmar

		Predictands		
		Maximum temperature	Minimum temperature	Precipitation
R^2	Cal	0.82	0.88	0.89
	Val	0.97	0.96	0.92
RMSE	Cal	0.51 °C	0.44 °C	37.19 mm
	Val	0.66 °C	0.84 °C	49.70 mm

Cal calibration, *Val* validation

Validation of the model on the other hand, was done independently for maximum, minimum temperature and precipitation with the observed data of 1981–1990 based on mean monthly value, standard deviation, monthly maximum value and time series plot of each simulated value are compared with corresponding observed data and results indicate a good agreement between observed data and simulated values.

5.4.2 Projection of Future Climate Variables

5.4.2.1 Projection of Future Temperature

The projected maximum temperature shows a decreasing trend whereas, minimum temperature tends to increase for the three scenarios and the three time windows considered (Table 5.7). The highest decrease in maximum temperature was 0.8 °C for 2020s time window in case of A2 scenario by HadCM3. On the contrary, the largest increase in minimum temperature has been observed in multiple cases with same absolute value of 0.4 °C for A2 scenario in 2020s and 2050s along with 2080s for A1B scenario.

The temperatures simulated by HadCM3 showed that although the increase in minimum temperature is not significant, it is however remarkable in the case of maximum temperature (−0.5 to −0.6 °C). It can be noted that in case of A2 and A1B scenarios, the maximum temperature simulated decreases abruptly in the near future (2020s) then tends to increase to the current temperature (32.7 °C). For maximum temperature, ECHAM5 shows a variation of −0.2 to 0 °C whereas a higher variability ranging from −0.8 to +0.1 °C is observed for HadCM3 from 2020s to 2080s. Similarly, for minimum temperature observed change for

Table 5.7 Future changes in maximum temperature (T_{max}) and minimum temperature (T_{min}) relative to baseline period (1961–1990) at Ngamoeyeik irrigation project, Myanmar

Scenarios		ECHAM5			HadCM3		
		2020s	2050s	2080s	2020s	2050s	2080s
Baseline T_{max} (°C)		32.7					
Baseline T_{min} (°C)		22.2					
Scenario A2	T_{max} (°C)	32.5	32.6	32.7	31.9	32.4	32.8
	Change (°C)	−0.2	−0.1	0	−0.8	−0.3	+0.1
	T_{min} (°C)	22.6	22.6	22.5	21.8	22.5	22.7
	Change (°C)	+0.4	+0.4	+0.3	−0.4	+0.3	+0.5
Scenario A1B	T_{max} (°C)	32.5	32.6	32.8	–	–	–
	Change (°C)	−0.2	−0.1	+0.1	–	–	–
	T_{min} (°C)	22.5	22.5	22.6	–	–	–
	Change (°C)	+0.3	+0.3	+0.4	–	–	–
Scenario B2	T_{max} (°C)	–	–	–	32.2	32.3	32.2
	Change (°C)	–	–	–	−0.5	−0.6	−0.5
	T_{min} (°C)	–	–	–	22.4	22.4	22.4
	Change (°C)	–	–	–	+0.2	+0.2	+0.2

ECHAM5 and HadCM3 ranges from +0.4 to +0.3 °C and −0.4 to +0.5 °C respectively for the corresponding time intervals. The analysis shows that the highest changes in both maximum and minimum temperature are observed in case of 2050s which has significant impact on IWR and rainfed paddy yield (discussed in later section).

5.4.2.2 Projection of Future Precipitation

A substantial increase in precipitation is observed in all time windows for A2 and B2 scenarios (Table 5.8). However, a considerable decline is noted from 2020s to 2080s in case of A1B scenario. A declining trend in precipitation is indicated by ECHAM5 for A1B scenario; similarly by both GCMs for A2 scenario whereas B2 indicates an increasing trend in precipitation magnitude. The highest magnitude of rainfall is observed in case of A2 scenario for 2020s (3,273 mm); for the expected abrupt increase in amount of rainfall within a short time, it is suggested to have a proper rainfall forecast throughout the study area to prevent harmful impacts. The increase in total amount of rainfall for A2 and B2 scenarios indicates reduced IWR. However, it is contradictory in case of A1B scenario (discussed later). Interestingly, minor differences can be noted in the projections for A2 scenario in ECHAM5 and HadCM3. However, both shows similar trend although the magnitude of increase for HadCM3 is relatively lower which has implication on IWR.

Table 5.8 Future changes in precipitation (Prcp) relative to baseline period (1961–1990) at Ngamoeyeik irrigation project, Myanmar

Scenarios		ECHAM5			HadCM3		
		2020s	2050s	2080s	2020s	2050s	2080s
Baseline T_{max} (°C)		32.7					
Baseline T_{min} (°C)		22.2					
Scenario A2	T_{max} (°C)	32.5	32.6	32.7	31.9	32.4	32.8
	Change (°C)	−0.2	−0.1	0	−0.8	−0.3	+0.1
	T_{min} (°C)	22.6	22.6	22.5	21.8	22.5	22.7
	Change (°C)	+0.4	+0.4	+0.3	−0.4	+0.3	+0.5
Scenario A1B	T_{max} (°C)	32.5	32.6	32.8	–	–	–
	Change (°C)	−0.2	−0.1	+0.1	–	–	–
	T_{min} (°C)	22.5	22.5	22.6	–	–	–
	Change (°C)	+0.3	+0.3	+0.4	–	–	–
Scenario B2	T_{max} (°C)	–	–	–	32.2	32.3	32.2
	Change (°C)	–	–	–	−0.5	−0.6	−0.5
	T_{min} (°C)	–	–	–	22.4	22.4	22.4
	Change (°C)	–	–	–	+0.2	+0.2	+0.2

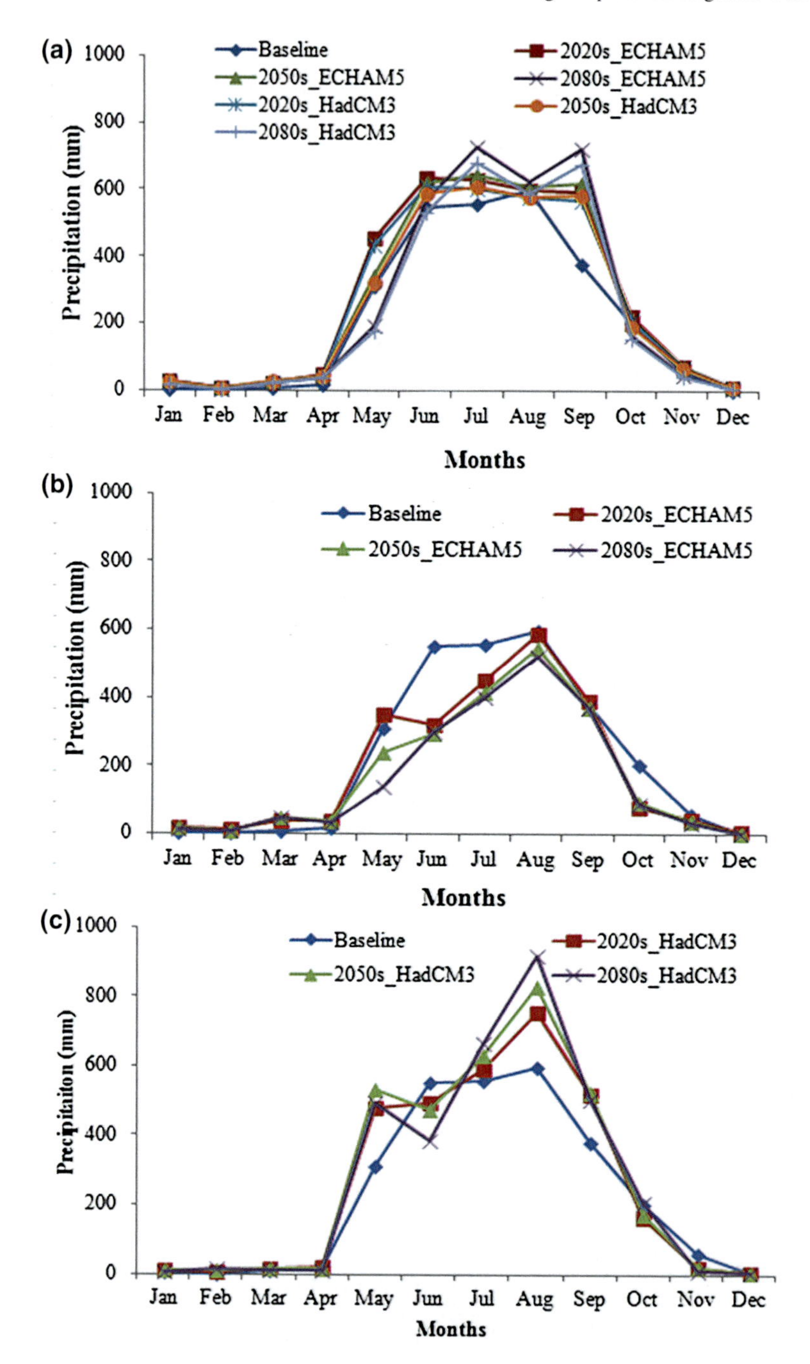

Fig. 5.4 Comparison of the forecasted precipitation with the baseline period (1961–1990) for **a** A2 **b** A1B and **c** B2 scenarios for the study area

Figure 5.4 illustrates the comparison of forecasted precipitation relative to the baseline period for the three scenarios and three time windows considered. Results show that forecasted precipitation does not deviate from the trend of the present condition. However, the magnitude is expected to change for all the three scenarios. In case of A2 scenario, simulations conducted from both climate models show the existence of double peaks (July and September) for 2080s in ECHAM5 and similar trend is also observed for 2020s by HadCM3 although with lower magnitude. The downscaling also suggest minor variation in the magnitudes of inter GCM outputs of monthly precipitation for all time windows in A2 scenario. The onset of the monsoon season follows the same trend for all the time windows in this case. However, prediction results for A1B scenario depict an early onset of monsoon with a lesser magnitude in June and July for the three future time windows (Fig. 5.4b). Although the offset is presumed to be at same time, there is however a reduction in the magnitude as compared to present for the three time windows. Forecasts of B2 scenario shows longer dry period and shorter wet period as the offset of monsoon is forecasted to be early for the three time windows. It can affect the reservoir operation since irrigation required for the paddy is regulated by precipitation pattern. An increasing trend of magnitude is also expected for August from 600 to 900 mm for 2080s which implies for rainfed rice it may be good since the cropping calendar indicates higher volume of water necessary for the development stage of irrigated paddy.

Figure 5.5 shows the percent change in the precipitation compared to baseline period (1961–1990) for the three time windows and three scenarios considered. Higher shifts are noticed in the dry seasons (December–April). Although the trend remains same for the three time windows but a significant change in magnitude has been observed. It is also interesting to note that for A2 scenario, downscaled climate data follows similar trend and minimum variation in magnitude of change for both ECHAM5 and HadCM3 GCMs. In A1B scenario, highest change is detected for 2020s in January which decreases to lowest in 2080s indicating probable high IWR for irrigated paddy in 2080s. An opposite trend is marked for B2 scenario with lowest change in precipitation for 2020s (February) which rises to maximum for 2080s (509 %).

5.4.3 Crop Model Set up

For setting up the crop growth model, the experimental data of daily weather, soil and field management collected from research center was used as the input parameter and the simulated results were compared with the observed data. The crop parameters were taken from the recommended default values by model guideline. The simulated data used to calibrate and validate the model are yield and biomass. AquaCrop is calibrated using the measured data sets in 2010 and the data set of 2011 was used for validation. The model has been calibrated for rainfed and irrigated conditions separately. The parameters used in model calibration are shown

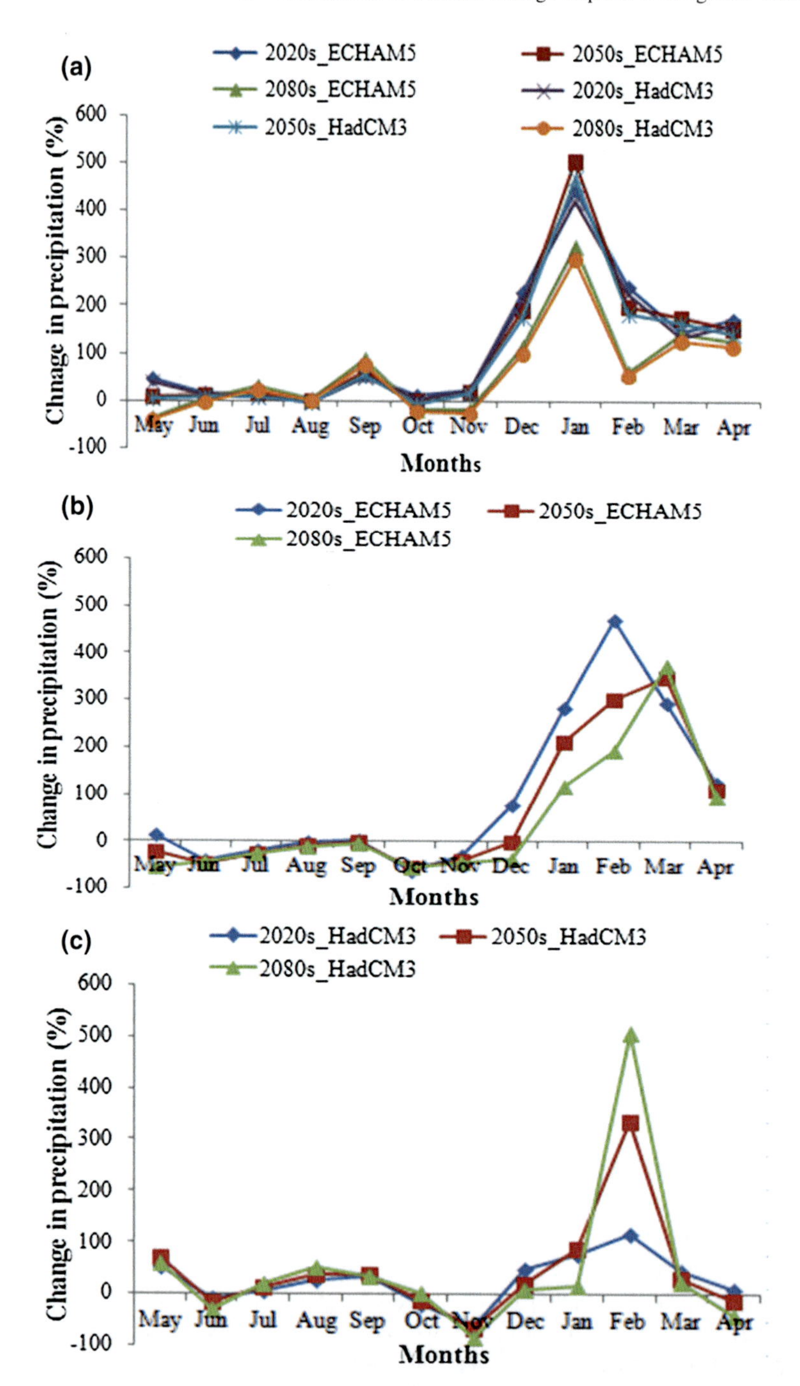

Fig. 5.5 Comparison of the forecasted precipitation (in terms of percentage) with baseline period (1961–1990) for **a** A2 **b** A1B and **c** B2 scenarios for the study area

Table 5.9 Calibrated parameters of AquaCrop model for rainfed and irrigated conditions

Description	Value		Units
	Rainfed	Irrigated	
Base temperature	8.2	8.0	°C
Cut-off temperature	31.2	29.0	°C
CC_0	1.1	1.8	%
CGC	8.0	8.3	%/day
CCx	90	95	%
CDC	8.1	8.4	%/day
Maximum effective rooting depth	0.38	0.4	m
Stomata stress coefficient (p_upper)	0.47	0.5	–
Senescence stress coefficient (p_upper)	0.50	0.55	–

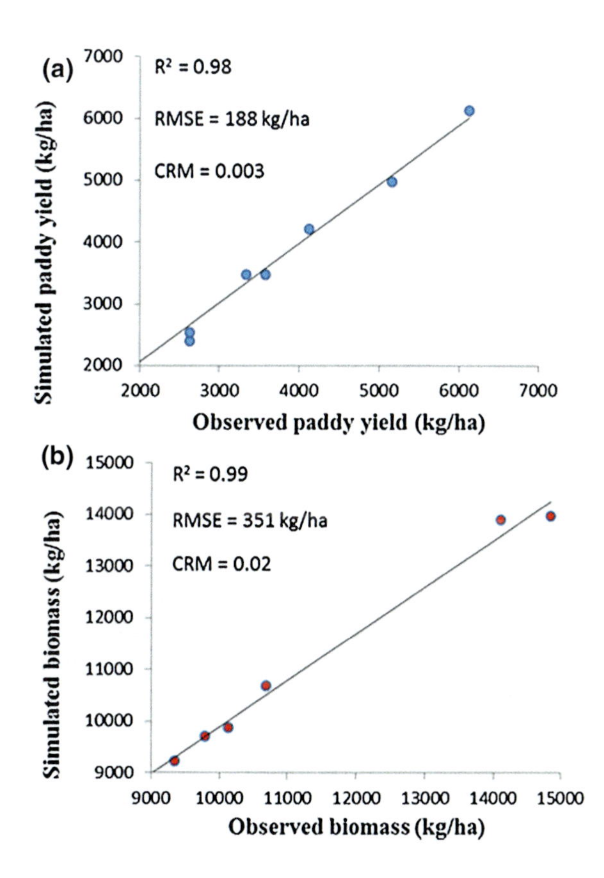

Fig. 5.6 Performance of AquaCrop model during calibration process for **a** paddy yield and **b** biomass

in Table 5.9 which include base temperature, cut-off temperature, initial canopy cover (Cco), canopy growth coefficient (CGC), maximum canopy (CCx), canopy decline coefficient (CDC), maximum effective rooting depth, stomata stress coefficient (upper) and senescence stress coefficient p(upper).

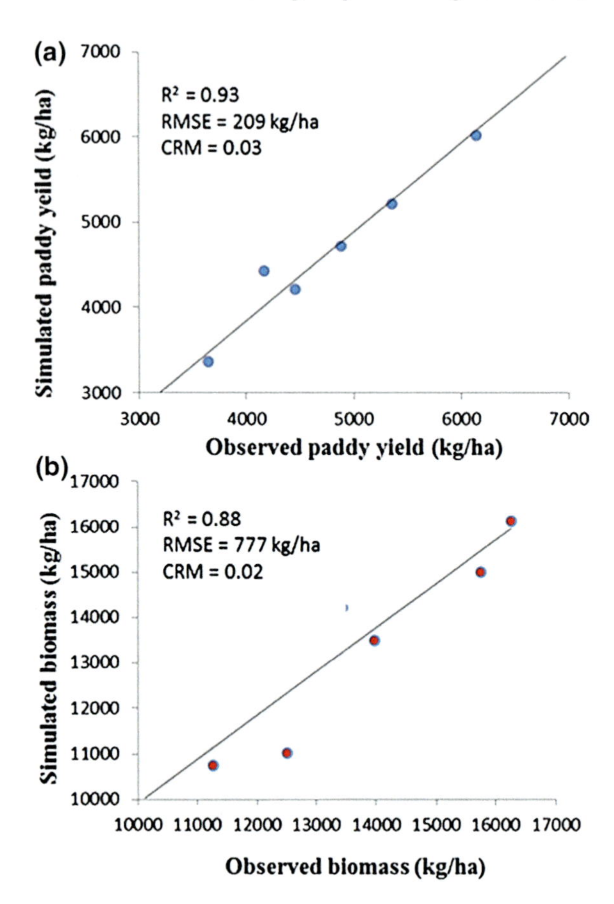

Fig. 5.7 Performance of AquaCrop model during validation process for **a** paddy yield and **b** biomass

Figures 5.6a, b and 5.7a, b represents the performance of AquaCrop model during calibration and validation for paddy yield and biomass respectively. Although a good relationship among the observed and simulated values can be inferred from the figures, CRM suggests however that the model overestimated both yield and biomass during the calibration and validation process. A higher RMSE of 209 and 777 kg/ha is observed in case of validation in contrast to 188 and 351 kg/ha for yield and biomass during calibration, respectively. Higher error in validation may be due to the contribution of climate and management factors leading to reduced observed yield and biomass. However the minimal existing error is still acceptable for our study.

Table 5.10 Future changes in irrigation water requirement relative to baseline period (1961–1990) for irrigated paddy in case of ECHAM5 (A2 and A1B scenarios) and HadCM3 (A2 and B2 scenarios) GCMs

Time period (GCM)	Allowable root zone depletion (%)	Initial condition	Baseline IWR (mm)	Scenario A2 (ECHAM5)		Scenario A2 (HadCM3)		Scenario A1B (ECHAM5)		Scenario B2 (HadCM3)	
				IWR	Change (mm)	IWR	Change (mm)	IWR	Change (mm)	IWR	Change (mm)
2020s	10	FC	527	486	−41	492	−35	474	−53	498	−29
		TAW 50 %	627	577	−50	601	−26	569	−58	582	−45
	25	FC	486	482	−4	478	−8	433	−53	475	−11
		TAW 50 %	607	527	−80	522	−85	547	−60	573	−34
2050s	10	FC	527	468	−59	460	−67	481	−46	479	−48
		TAW 50 %	627	551	−76	546	−81	572	−55	568	−59
	25	FC	486	431	−55	433	−53	448	−38	435	−51
		TAW 50 %	607	524	−83	523	−84	552	−55	545	−62
2080s	10	FC	527	498	−29	504	−23	491	−36	469	−58
		TAW 50 %	627	589	−38	577	−50	577	−50	556	−71
	25	FC	486	472	−14	480	−6	460	−26	425	−61
		TAW 50 %	607	569	−38	602	−5	553	−54	532	−75

TAW total available water, *FC* field capacity

5.4.4 Projection of Future Irrigation Water Requirement (IWR)

The future IWR for the irrigated paddy in case of future time windows and three scenarios was simulated by AquaCrop. Two plausible initial soil moisture conditions combined with two assumed allowable root zone depletion conditions namely at Field Capacity (FC) and 50 % of Total Available Water (TAW) were considered for the simulation (Table 5.10). The nutrient supply is presumed to be at optimum application rate as that of the current situation. The agricultural management practices are also assumed to be unaltered as compared to the present conditions.

Compared to present condition, a decrease in the IWR is observed for the three scenarios at all-time windows considered. At FC, lesser IWR is observed as compared to presumed initial condition of TAW at 50 % for all allowed root zone depletion conditions in the time windows considered. It is also evident that, with increase in allowable root zone depletion, there is a substantial reduction in IWR. Simulation done for ECHAM5 in case of A2 scenario suggests more irrigation water requirement in terms of depth (mm) for 2080s followed by 2020s and 2050s. A similar trend is observed in terms of magnitude for A1B scenario, where highest IWR is observed in 2080s for all the conditions followed by 2050s and 2020s. However, in the case of B2 scenario, IWR reduces from 2020s to 2080s. The maximum reduction in IWR (83 mm) as compared to present condition is observed for 2050s with 25 % allowable root zone depletion and 50 % TAW initial condition. Simulation for IWR done with the outputs of HadCM3 (A2 scenario) validates the higher degree of uncertainty in 2080s when compared to that of ECHAM5 projections. IWR simulation results obtained from ECHAM5 for 25 % allowable root zone depletion at FC and TAW 50 %, suggest a reduction of 14 and 38 mm respectively whereas in case of HadCM3 projections IWR stands at 6 and 5 mm for the corresponding conditions.

The trend in forecasted precipitation and minimum temperature can be attributed to the observed decreasing trend in IWR for the three scenarios. It can be clearly observed that for any scenario at a particular time window with projected high precipitation, a lower IWR persists. For instance in case of A2 scenario (ECHAM5), for 2020s, 2050s and 2080s the respective precipitation are 194.8 mm, 194.3 mm and 146.2 mm and the corresponding IWR for 10 % allowable root zone depletion and initial condition at FC are 486, 468 and 498 mm. Similar results were shown by Gerten et al. (2011) and Olesen et al. (2007) which indicates reduction in IWR by 4 to 82 % by the end of 21st century varying on crop and location. The projected decrease in IWR for the future climate scenarios indicates the need of better management plan for the diversion of the reservoir water to other sectors during the dry months.

Table 5.11 Future changes in rainfed rice yield relative to baseline period (1961–1990) for ECHAM5 (A2 and A1B scenarios) and HadCM3 (A2 and B2 scenarios) GCMs

Scenarios	Baseline yield (t/ha)	2020s		2050s		2080s	
		Yield (t/ha)	Changes (%)	Yield (t/ha)	Changes (%)	Yield (t/ha)	Changes (%)
A2_ECHAM5	2.965	3.590	+21.1	3.932	+32.6	4.159	+40.3
A2_HadCM3		3.612	+21.8	3.991	+34.6	4.029	+35.9
A1B_ECHAM5		3.573	+20.5	3.783	+27.6	3.745	+26.3
B2_ECHAM5		3.571	+20.4	3.465	+16.9	3.455	+16.5

5.4.5 Projection of Future Rainfed Paddy Yield

An increase in the rainfed paddy yield is observed for the three scenarios and the three time windows considered. An explicit increasing trend is observed from 2020s to 2080s for A2 scenario in case of both GCMs. The projected yield for ECHAM5 shows a change of +40.3 % for A2 scenario whereas for HadCM3 +35.9 % change is observed for the corresponding scenario. It can also be noted that the uncertainty is higher in magnitude for the late part of century relative to the early and mid-part. In addition, a fluctuating trend is observed in case of A1B and a decreasing trend is observed in B2 scenario (Table 5.11). Maximum variation is observed for A2 scenario ranging from +21.1 to +40.3 % from 2020s to 2080s. It can be observed that for rainfed paddy, the trend of yield is irrespective of the trend of maximum temperature and precipitation. However, the increase in yield follows the same trend as that of minimum temperature. Modeled higher yields in case of A2 followed by A1B and B2 scenarios may also be induced due to the accumulated CO_2 concentration in the atmosphere. Higher CO_2 level causes increase in minimum temperature which combined together due to heat-induced spikelet aggravated growth and increased biomass and subsequent grain yield (Wassmann 2007; Krishnan et al. 2007). Similar increase in yield was also shown by Alexandrov et al. (2002) and Laux et al. (2010) for Austria and Cameroon respectively. In case of Southeast Asia, Northeast Thailand and winter-spring cropping pattern of Vietnam are supposed to be benefitted by climate change as modeling study suggests increased rice yield (SeaStart 2006).

The decreasing trend of IWR for summer paddy is observed for all three time windows. Since the paddy is already in irrigated condition and the maximum yield is attained, the paddy doesn't respond to further increase in rainfall during summer season. Therefore very little increase or no further increase in the yield of summer paddy can be expected.

The projected increase in yield of rainfed paddy due to climate change is an indication of increased food security in the study area. However, evaluation of adaptation strategies can be performed in order to enhance the yield up to its potential level. Although it is expected to have an increased paddy yield under climate change but severe risk exists due to the increased precipitation as high magnitude of precipitation can also cause unexpected floods in the region if proper management practices are not being taken.

5.5 Conclusions

This study is divided into three parts: first is forecasting the future climate variables, then using the projected climate variables to forecast the future irrigation water requirements for irrigated paddy and thirdly, assessing the impacts on future rainfed rice productivity. SDSM has been used to downscale the coarse resolution of climate variables from the GCMs (HadCM3 and ECHAM5). A2, A1B and B2 scenarios were used to assess the IWR and rainfed rice yield. SDSM was calibrated for the period of 1961–1990 and validated for 1991–2000; performance statistics shows modeled outputs were in good agreement with observed ones. A decreasing trend in maximum temperature is observed for A2 scenario for the projections by ECHAM5 whereas a contradictory increase is observed for the projection by HadCM3. A fluctuation in forecasted temperature is noted for A1B and B2 scenarios for the three time windows considered (2020s, 2050s and 2080s). In case of minimum temperature, an increasing trend is observed for A2 and A1B scenarios however, an increase of +2 °C is expected to prevail constantly for B2 scenario for the three time windows. The future precipitation is expected to increase in magnitude by 14.4 and 19.7 % for A2 and B2 scenarios by 2080s compared to present. However, a decreasing trend is observed in case of A1B scenario which is expected to reduce more in future time intervals (−28.7 % in 2080s compared to present).

AquaCrop model was used for the study area and was calibrated and validated based on the field experimental data acquired for the agricultural research center. Projection of future IWR was done for the irrigated paddy based on the downscaled climate data. Due to the forecasted increase in the winter precipitation, the IWR is expected to be lowered by the end of the century for A2 and B2 scenarios. However, even the precipitation is forecasted to decrease, in the case of A1B scenario, the IWR is expected to be reduced due to shift in temperature pattern. Results obtained for rainfed rice simulated by AquaCrop suggests an explicit trend of increasing yield in the region for A2 scenario. A fluctuating increased trend is observed for the A1B scenario and reducing trend is noted for B2 scenario (+20.4 % in 2020s to +16.5 % in 2080s). It can also be inferred that due to the reduced IWR in the future the reservoir operation needs to be evaluated to divert the water to other sectors for better water resources management. The results can be further utilized to investigate the effects of different cropping patterns with varying crop calendars on the water demand.

References

Abedinpour M, Sarangi A, Rajput TBS, Singh M, Pathak H, Ahmad T (2012) Performance evaluation of AquaCrop model for maize crop in a semi-arid environment. Agric Water Manag 110:55–56

ADB (2009) The economics of climate change in Southeast Asia: a regional review. Report by Asian Development Bank. Asian Development Bank, Jakarta

Alcamo J, Dronin N, Endejan M, Golubev G, Kirilenko A (2007) A new assessment of climate change impacts on food production shortfalls and water availability in Russia. Glob Environ Change 17:429–444

Alexandrov V, Eitinger J, Cajic V, Oberforster M (2002) Potential impact of climate change on selected agricultural crops in north-eastern Austria. Glob Change Biol 8(4):372–389

Allen RG, Pereira LS, Raes D, Smith M (1998) Crop evapotranspiration-guidelines for computing crop water requirements. FAO irrigation and drainage paper 56. FAO, Rome

Amthor J (1998) Perspective on the relative insignificance of increasing atmospheric CO_2 concentration to crop yield. Field Crops Res 58:109–127

Artlert K, Chaleeraktrakoon C, Nguyen V (2013) Modeling and analysis of rainfall processes in the context of climate change for Mekong, Chi and Mun river basin (Thailand). J Hydro-environ Res 7:2–17

Babel MS, Agarwal A, Swain DK, Herath S (2011) Evaluation of climate change impacts and adaptation measures for rice cultivation in Northeast Thailand. Clim Res 46:137–146

Bárdossy A, Plate EJ (1992) Space-time model of daily rainfall using atmospheric circulation pattern. Water Resour Res 28(5):1247–1259

De Silva CS, Weatherhead EK, Knox JW, Rodriguez-Diaz JA (2007) Predicting the impacts of climate change-A case study of paddy irrigation water requirements in Sri Lanka. Agric Water Manage 93:19–29

Easterling WE, Apps M (2005) Assessing the consequences of climate change for food and forest resources: a view from the IPCC. Clim Change 70:165–189

Erda L, Wei X, Yinlong X, Yue L, Liping B, Liyong X (2005) Climate change impacts on crop yield and quality with CO_2 fertilization in China. Philos Trans R Soc B 360:2149–2154

FAO (1998) Crop evapotranspiration—guidelines for computing crop water requirements, FAO Irrigation and drainage paper 56, Food and Agriculture Organization of the United Nations, Viale delle Terme di Caracalla, Rome, Italy

FAO (2009) FAO/WFP crop and food security assessment mission to Myanmar. Special report by Food and Agriculture Organization. Food and Agriculture Organization, Rome

Fuhrer J (2003) Agroecosystem responses to combinations of elevated CO_2, ozone and global climate change. Agric Ecosyst Environ 97:1–20

Gerten D, Heinke J, Hoff H, Biemans H, Fader M, Waha K (2011) Global water availability and requirements for future food production. J Hydrometeor 12:885–899

Greets S, Raes D, Garcia M, Miranda R, Cusicanqui JA, Taboada C, Mendoza J, Huanca R, Mamani A, Condori O, Mamani J, Morales B, Osco V, Steduto P (2009) Simulating yield response of Quinoa to water availability with AquaCrop. Agron J 101:499–508

Horie T (2005) Climate change and food security with special attention to rice. In: Proceedings of global environ action (GEA) conference for sustainable future 2005, Tokyo

IFPRI (2009) Climate change impact on agriculture and costs of adaptation. Food policy report of International Food Policy Research Institute. International Food Policy Research Institute, Washington, D.C.

Inthavon T, Jintrawet A, Chinvanno S, Snidvongs A (2004) Impact of climate change on rainfed lowland rice production in Savannakhet Provience, Lao PDR. Report of APN CAPABLE project, Southeast Asia START Regional Center, Chulalongkorn University, Bangkok

IPCC (2007) Climate change: impacts, adaptation and vulnerability. Contribution of working group II to the 4th assessment report of the intergovernmental panel on climate change. Intergovernmental Panel on Climate Change. Cambridge University Press, Cambridge

Kang Y, Khan S, Ma X (2009) Climate change impacts on crop yield, crop water productivity and food security—a review. Prog Nat Sci 19:1665–1674

Krishnan P, Swain DK, Bhaskar BC, Nayak SK, Dash RN (2007) Impact of elevated CO_2 and temperature on rice yield and methods of adaptation as evaluated by crop simulation studies. Agric Ecosyst Environ 122:233–242

Laux P, Jäckel G, Tingem RM, Kunstmann H (2010) Impact of climate change on agricultural productivity under rainfed conditions in Cameroon—a method to improve attainable crop yield by planting date adaptations. Agric For Meteorol 150:1258–1271

Long SP, Ainsworth EA, Leakey ADB, Nösberger J, Ort DR (2006) Food for thought: lower-than-expected crop yield stimulation with rising CO_2 concentrations. Science 312:1918–1921

Maeda EE, Pellikka PKE, Clark BJF, Siljander M (2011) Prospective changes in irrigation water requirements caused by agricultural expansion and climate changes in the eastern arc mountains of Kenya. J Environ Manag 92:982–993

Mkhabela MS, Bullock PR (2012) Performance of the FAO AquaCrop model for wheat grain yield and soil moisture simulation in Western Canada. Agric Water Manag 110:16–24

Molua EL (2009) An empirical assessment of the impact of climate change on smallholder agriculture in Cameroon. Glob Planet Change 67:205–208

Nuorteva P, Keskinen M, Varis O (2010) Water, livelihoods and climate change adaptation in the Tonle Sap Lake area, Cambodia: learning for the past to understand the future. J Water Clim Change 1(01):87–101

Olesen JE, Jensen T, Petersen J (2000) Sensitivity of field-scale winter wheat production in Denmark to climate variability and climate change. Clim Res 15:221–238

Olesen JE, Carter TR, Díaz-Ambrona CH, Fronzek S, Heidmann T, Hicker T, Holt T, Minguez MI, Morales T, Palutikof JP, Quemada M, Ruiz-Ramos M, Rubæk GH, Sau F, Smith B, Sykes MT (2007) Uncertainties in projected impacts of climate change on European agriculture and terrestrial ecosystems based on scenarios from regional climate models. Clim Change 81:123–143

Raes D, Steduto P, Hsiao TC, Fereres E (2009a) AquaCrop—the FAO crop model to simulate yield response to water: II. Main algorithms and software description. Agron J 101:438–447

Raes D, Steduto P, Hsiao TC, Fereres E (2009b) Crop water productivity. Calculation procedures and calibration guidance. AquaCrop version 3.0. FAO, Land and Water Development Division, Rome

Russo JM, Zack JW (1997) Downscaling GCM output with a mesoscale model. J Environ Manag 49:19–29

SeaStart (2006) Southeast Asia regional Vulnerability to changing water resource and extreme hydrological events due to climate change. Final technical report of Southeast Asia START regional center technical report no. 15. Bangkok, Thailand

Shrestha S, Gyawali B, Bhattarai U (2013) Impacts of climate change on irrigation water requirements for rice-wheat cultivation in Bagmati River basin, Nepal. J Water Clim Change (in press)

Smith JB, Hulme M (1998) Climate change scenarios (Chapter 3). In: Feenstra J, Burton I, Smith JB, Tol RJS (eds) Handbook on methods of climate change impacts and adaptation strategies. UNEP/IES, Version 2.0, Amsterdam, The Netherlands

Steduto P, Hsiao TC, Raes D, Fereres E (2009) AquaCrop—the FAO crop model to simulate yield response to water: I. Concepts and underlying principles. Agron J 101(3):426–437

UNDP (2011) Integrated household living conditions survey in Myanmar. Poverty profile report of Myanmar. United Nations Development Program, Yangoon

USDA (1970) Irrigation water requirements. Technical release no. 21. USDA Soil Conservation Service, Washington, DC

Wassmann R (2007) Coping with climate change. International Rice Research Institute, Manila. http://beta.irri.org/news/images/stories/ricetoday/63/feature_coping%20with%20climate%20change.pdf. Accessed 12 November 2012

Wilby RL, Dawson CW, Barrow EM (2002) SDSM—a decision support tool for the assessment of regional climate change impacts. Environ Model Softw 17:147–159

Yano T, Aydin M, Haraguchi T (2007) Impact of climate change on irrigation demand and crop growth in a Mediterranean environment of Turkey. Sensors 7:2297–2315

Young KB, Cramer GL, Wailes EJ (1998) An economic assessment of the Myanmar rice sector: current developments and prospects. Report of Arkansas global rice project. Arkansas Agricultural Experiment Station, Arkansas

Chapter 6
Adaptation Strategies for Rice Cultivation Under Climate Change in Central Vietnam

Abstract This study investigates the impact of climate change on winter and summer rice (*Oryza sativa*) yield and evaluates several adaptation measures to overcome the negative impact of climate change on rice production in the Quang Nam province of Vietnam. Future climate change scenarios for time periods in the 2020s, 2050s and 2080s were projected by downscaling the outputs of the General Circulation Model (GCM), Hadley Centre Coupled Model, version 3 (HadCM3) A2 and B2 scenarios. The AquaCrop model was used to simulate the impact of future climates on rice yield. The minimum and maximum temperature of the province is projected to increase by 0.35–1.72 °C and 0.93–3.69 °C respectively in future. Similarly, the annual precipitation is expected to increase by 9.75 % in the 2080s. Results show that climate change will reduce rice yield from 1.29 to 23.05 % during the winter season for both scenarios and all time periods, whereas an increase in yield by 2.07–6.66 % is expected in the summer season for the 2020s and 2050s; relative to baseline yield. The overall decrease of rice yield in the winter season can be offset, and rice yield in the summer season can be enhanced to potential levels by altering the transplanting dates and by introducing supplementary irrigation. Late transplanting of rice shows an increase of yield by 20–27 % in future. Whereas supplementary irrigation of rice in the winter season shows an increase in yield of up to 42 % in future. Increasing the fertilizer application rate enhances the yield from 0.3 to 29.8 % under future climates. Similarly, changing the number of doses of fertilizer application increased rice yield by 1.8–5.1 %, relative to the current practice of single dose application. Shifting to other heat tolerant varieties also increased the rice production. Based on the findings, changing planting dates, supplementary irrigation, proper nutrient management and adopting to new rice cultivars can be beneficial for the adaptation of rice cultivation under climate change scenarios in central Vietnam.

Keywords Agro-adaptation · AquaCrop · Climate change · Rice · Vietnam

S. Shrestha, *Climate Change Impacts and Adaptation in Water Resources and Water Use Sectors*, Springer Water, DOI 10.1007/978-3-319-09746-6_6

6.1 Introduction

The Fifth Assessment Report of the United Nations Intergovernmental Panel on Climate Change (IPCC) reported that the future greenhouse gas emission will keep on rising, and the global average temperature is likely to be increased from 0.3 to 4.8 °C, based on various scenarios (Stocker et al. 2013). Vietnam, without exception, has already experienced an average increase in temperature of 0.5–0.7 °C and an average reduction in rainfall of 2 % in the last five decades (MONRE 2009). This kind of change in the climatic variables has already affected the crop growth patterns and reduced yield in many parts of the world (Ray et al. 2012). Higher temperatures can potentially affect the physiological processes such as photosynthesis and respiration (Yang and Zhang 2006). In regions where the temperature limits the length of the growing season, a warmer climate is beneficial for crop yields (Meza et al. 2008). Rising temperatures can cut down the growth, grain filling rate and duration of crop maturity (Boote 2011).

Vietnam is the country most vulnerable to climate change (Dasgupta et al. 2007) where rice (*Oryza sativa*) cultivation accounts for more than three-quarters of the country's total annual harvested agricultural area and employs about two-thirds of the rural labour force, thus making a significant contribution to rural livelihoods (Nguyen 2006; Vu and Glewwe 2008). Currently Vietnam stands as the second-largest (after Thailand) exporter worldwide and the world's seventh-largest consumer of rice (FAO 2010). Therefore, rice production in Vietnam makes a significant contribution to global food security. According to statistics from the Ministry of Agriculture and Rural Development (MARD) in 2009, the area of agricultural planted land is about 9.4 million hectares, including 4 million hectares of rice land, and the figure targets about 10 million hectares in 2020. However several challenges, such as low soil fertility, salinity intrusion, insect pest infestation, limit increasing rice productivity in Vietnam, and climate change is the additional factor that will add uncertainty in rice production. Studies for the Southeast Asian region show that climate change could lower agricultural productivity by 15–26 % in Thailand, 2–15 % in Vietnam, 12–23 % in the Philippines, and 6–18 % in Indonesia (Zhai and Zhuang 2009). Nguyen et al. (2008) found that the Mekong River Delta and the coastal areas in the North of the central region are most vulnerable to the impact of global warming in Vietnam due to rising sea levels.

The potential impact of climate change on rice productivity is reported in many recent studies such as Babel et al. (2011), Luo et al. (2013) and Soora et al. (2013). Although climatic variables are uncontrollable, factors such as cultivars, soil, water and nutrients can be managed in order to counteract the adverse effects of climate change (Moradi et al. 2013). Estimating the impact of climate change on crop yield and the evaluation of appropriate adaptation and mitigation strategies are of extreme concern (Jalota et al. 2012; Dharmarathna et al. 2014) to either stabilize or improve the crop yields. Several studies at various places have confirmed that rice cultivation without considering proper adaptation and mitigation strategies is problematic (Adejuwon 2006; Wassmann et al. 2009; Iizumi et al. 2011; Tao et al. 2012; Poudel and Kotani 2013). Recent literature has also reported the adequateness of proper agronomic adaptation strategies in relation to climate-induced yield losses in different regions (Tingem and Rivington 2009; Chhetri and Easterling 2010; Gouache et al. 2012; Mishra et al. 2013).

Adaptation strategies in agriculture to offset the negative impact of climate change are found to be shifting planting and transplanting dates, changing sowing density (Bindi and Olesen 2000), irrigation management, development of new agricultural areas and use of heat resistant varieties (Rosenzweig and Tubiello 2007; Babel and Turyatunga 2014). Babel et al. (2011) evaluated the impact of climate change on rice production in Thailand and emphasized the need for effective adaptation to stabilize the rice yield in future. A study on the super-ensemble based probabilistic projection approach was applied to project maize productivity and water use in the North China plain by Tao and Zhang (2010). Results revealed that early and late planting of temperature sensitive and high-temperature tolerant varieties respectively, are suitable for effective adaptation. A considerable number of studies have also validated that the suitable adaptation options are usually region specific and need to be appraised as per the location (Bryan et al. 2009; Tao and Zhang 2010).

The aims of the present study are: (a) to forecast the future climate using the output of the global circulation model; (b) to estimate the impact of future climates on rice yield; and (c) to explore the possibilities of employing different adaptation measures to offset the negative impact of climate change on rice production in the Quang Nam province of Vietnam.

6.2 Material and Methods

6.2.1 Study Area Description

Quang Nam is a coastal province in Central Vietnam with a land area of 10,438 km^2 where hills account for 72 % of the land area. Geographically it lies between latitude 14°58′–16°18′N and longitude 107°08′–108°47′E (Fig. 6.1). Higher elevation persists in the Western border towards Laos whereas the Eastern region forms plains and thus rice is cultivated. More than 50 % of the total land area in this province is covered by forest which belongs to higher altitudes. However, agriculture is generally practiced in the Eastern region of the country with a total area of 1,119 km^2 of which rice production contributes 75 %.

The climate of the region is tropical where the maximum and minimum temperature ranges from 24.8 to 34.9 and 19.1 to 25.6 °C respectively. The average annual rainfall is 2,700 mm with 80 % occurring from August to January, forcing the farmers to opt for rainfed rice cultivation in winter. However, due to the presence of high hills in the Western region, year around water is available in perennial streams, allowing the farmers to additionally cultivate in the dry season.

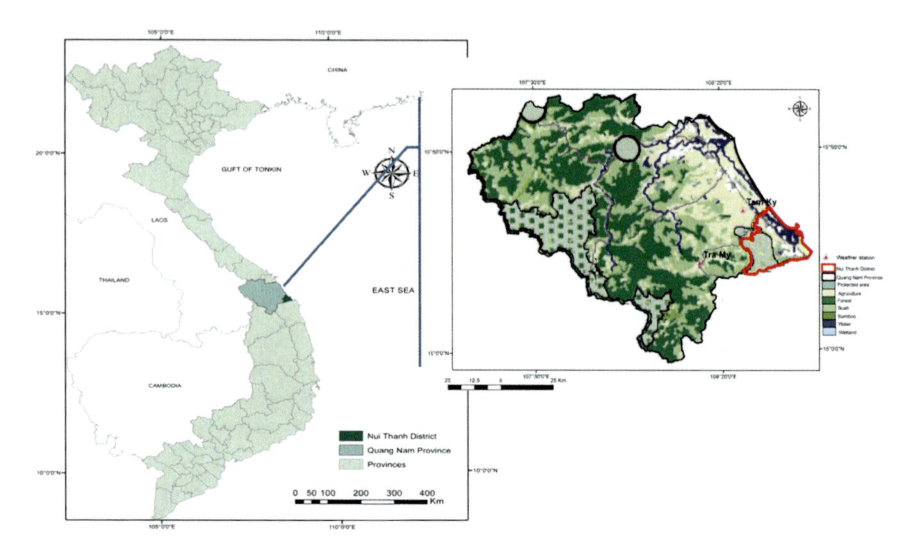

Fig. 6.1 The location map of the study area in Vietnam

The soil texture throughout the root zone is silty clay, and can be classified as entisols with low organic matter and neutral pH. The bulk density of the soil suggests it has a high ability to hold water and thus deep percolation of water is not a problem in the site.

Due to the short growing period of the rice cultivars, a double cropping system is generally practiced in the region, with winter cropping (rainfed) from the third week of January to the fourth week of April, followed by the second season summer cropping (irrigated) starting from the third week of May to the fourth week of August. The intermittent showers due to the northwest monsoon during the crop growth period from January to May serve to provide the required water during the rainfed rice cultivation.

6.2.2 Data Collection

The meteorological data was collected for the Tra My weather station (15°19′N and 108°13′E) where the field experiments for rice were conducted by the Quang Nam Department of Agriculture and Rural Development, Agricultural Division in the Nui Thanh district. The daily weather data consists of precipitation, sunshine hours, wind speed, relative humidity, and maximum and minimum temperature for the period of 1961–2010. The climate variables (precipitation and maximum, minimum temperature) for the future time periods were obtained from the Global Circulation Model (GCM), Hadley Centre Coupled Model, version 3 (HadCM3) developed by the Met Office Hadley Centre, England (www.metoffice.gov.uk). The Intergovernmental Panel on Climate Change (IPCC) has defined standard greenhouse gas emission scenarios for use in the evaluation of projected climate change based on various socio-economic, energy use and technological advancement (IPCC 2007). The A2 scenario, with an estimated carbon dioxide concentration of $\sim$860 ppm, postulates extreme heterogeneous world conditions with a slowly increasing population rate, regionally oriented economic development, self-reliant governance and slower technological development (IPCC 2000). The assumptions underlying the B2 scenario, with an estimated carbon dioxide concentration of $\sim$550 ppm, are local and regional governance solutions for economic, equality and environmental sustainability. Population growth is assumed to be less relative to the A2 scenario, and technological development is also less rapid. Economic growth is considered intermediate and environmentally protected with social equity at regional and local levels (IPCC 2000). For Vietnam, a fast developing nation, industrial growth is expected to be very rapid in the near future which can lead to an accumulation of high concentration greenhouse gases (GHGs) by the end of the century and therefore the A2 scenario (extreme scenario) was considered. Also, another expectation is that in the future people may prioritize environment sustainability and place more emphasis on local solutions to economic and social perspectives, leading to less industrial growth, and hence the B2 scenario was chosen for this study. Moreover, several studies on climate change impact assessment in this region have considered these two scenarios and given them primary importance as they are very realistic in this region (Babel et al. 2011; Mainuddin et al. 2011, 2013; Shrestha et al. 2014). In addition, HadCM3 was selected based on the findings of IPCC (2007) and Ruosteenoja et al. (2003), where they described the ability of representing the present day climatic data by HadCM3 compared to other models in Southeast Asia including Vietnam.

The details of field experiments were obtained from the Quang Nam Department of Agriculture and Rural Development, Agricultural Division in the Nui Thanh district. The crop data comprising transplanting dates, flowering dates, maturity dates, and irrigation schedule with amounts and rice yield was collected for rice cultivars, namely CH207, TBR1 and OM6162 which are grown in both summer and winter seasons in the study area. Although the details of three cultivars are provided, the farmers mostly prefer to grow CH207 due to the lower investment cost associated with it in terms of management. Table 6.1 represents the characteristics of the two rice cultivars considered in this study. Table 6.2 illustrates the details of the secondary data collected for field trials conducted in the study area under controlled conditions. The data is presented for both rainfed and irrigated rice field trials which were performed from 2001 to 2010, laid in a randomized complete block design. Fertilizer was applied at the recommended dose of 100 kg/ha of Ammonium (NH_4-N), 50 kg/ha Phosporous pentaoxide (P_2O_5) and 50 kg/ha of Potassium oxide (K_2O) in two splits; the first 10 days prior to transplantation and the second at 30 days after transplantation (DAT). Irrigation water for the field trials during the summer cropping was provided by a flooding method supplied by channels. Soil bunds were provided to retain the water in the transplanted region. The seedlings were transplanted at a depth of 2 cm and spacing of 20 $\times$ 20 cm was provided for each plant.

Table 6.1 Characteristics of CH207, TBR1 and OM6162 rice cultivars cultivated in the Quang Nam province, Vietnam

Characteristics	Name of rice varieties		
	CH207	TBR1	OM6162
Days of maturity	Winter crop: 110–115	Winter crop: 110–120	Winter crop: 92–105
	Summer crop: 95–100	Summer crop: 100–105	Summer crop: 90–100
Plant height (cm)	97–100	95–105	98–109
Length of panicle (cm)	22–24	20–25	21–26
Grain type	Medium	Long	Long
Grain form	Regular milled white rice	Regular milled white rice	Regular milled white rice
Grains per panicle	120	125	128
1,000 grains weight (gram)	26–27	25–28	27–30
Yield (t/ha)	In precarious conditions: 4.5–5.5	In precarious conditions: 5.0–6.0	In precarious conditions: 6.1–6.8
	In stable conditions and enough water: 5.7–6.5	In stable conditions and enough water: 6.5–7.5	In stable conditions and enough water: 6.9–7.8

Table 6.2 Crop growth characteristics for winter (rainfed) and summer (irrigated) rice used in calibration and validation of AquaCrop

Year	Primary tillage	Harrowing	Transplanting date	Flowering (Anthesis)	Maturity	Grain yield (t/ha)	Biomass yield (t/ha)
Winter (Rainfed)							
2001	3 Jan	15 Jan	21 Jan	4 Apr	5 May	4.72	14.66
2002	4 Jan	13 Jan	25 Jan	5 Apr	30 Apr	5.04	15.21
2004	10 Jan	24 Jan	25 Jan	12 Apr	8 May	5.27	14.98
2007	28 Dec	10 Jan	22 Jan	7 Apr	3 May	4.81	14.79
2009	26 Dec	4 Jan	12 Jan	8 Apr	3 May	5.21	14.89
Summer (Irrigated)							
2003	26 Apr	3 May	12 May	14 Jul	17 Aug	5.38	15.65
2005	10 May	16 May	26 May	4 Aug	7 Sep	5.31	15.26
2006	28 Apr	5 May	14 May	18 Jul	24 Aug	5.22	15.02
2008	5 May	14 May	22 May	24 Jul	26 Aug	5.18	15.23
2010	28 Apr	5 May	13 May	18 Jul	20 Aug	5.51	15.52

Source Quang Nam Department of Agriculture and Rural Development

Soil data was collected from the Quang Nam State Land and Development section. The data obtained includes depth-wise soil texture, pH, phosphorous, nitrogen, carbon content, field capacity and permanent wilting point of the soil.

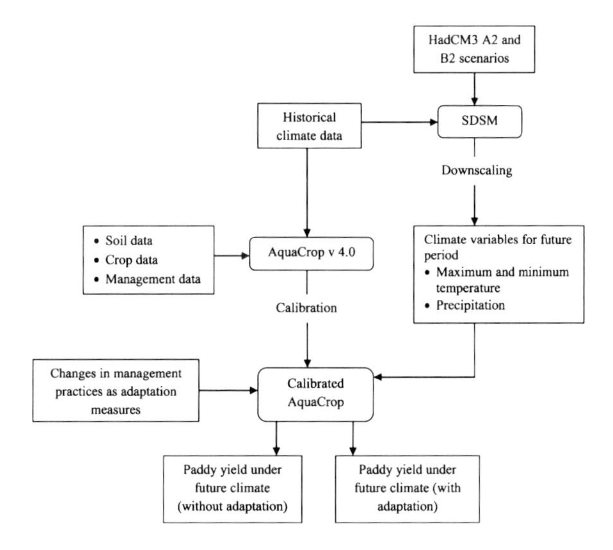

Fig. 6.2 The methodological framework of the study at the Quang Nam province, Vietnam

The methodological flowchart followed in this study is shown in Fig. 6.2. Firstly, the future climate data was downscaled to province by using the Statistical Downscaling Model (SDSM) followed by setting up of the crop model AquaCrop, based on the historical climatic data and agronomic management details. Due to the availability of separate field experimental data for winter and summer cropping as in Table 6.2, the AquaCrop model was calibrated based on 5 year splits of yield and biomass for the corresponding seasons. In addition, the calibrated model was used to simulate the future climate change impact on rice yield for winter and summer rice based on the future downscaled climate variables (daily maximum, minimum temperature and daily precipitation) for A2 and B2 scenarios as input into AquaCrop for three time periods: 2014–2040 (2020s), 2041–2070 (2050s) and 2071–2090 (2080s). For future time periods, predetermined CO_2 concentration integrated in the model was used for projection of the rice yield. Furthermore, several agro-adaptation measures were evaluated to offset the negative impact of climate change.

6.2.3 Generation of Climate Change Scenarios

The climate variables of HadCM3 GCM have coarse spatial resolution, 2.5° latitude by 3.75° longitude (CICS 2012) and hence it is unsuitable to use the outputs for the province scale study. Therefore, it is necessary to downscale the climate variables at the province level for impact studies (Giorgi and Mearns 1991). The Statistical DownScaling Model (SDSM) was used to downscale the climate variables at the province level. The SDSM develops transfer functions among large scale predictor variables and station level climate variables. In case of precipitation, the model additionally uses stochastic techniques to artificially inflate the variance of the downscaled daily time series to better accord with the observations (Wilby and Dawson 2007).

The SDSM model includes seven steps for the entire downscaling process: data quality control and transformation, screening of predictor variables, model calibration, weather generation, statistical analysis, graphing climate data along model outputs and scenario generation. The National Centres for Environmental Prediction (NCEP) regional scale predictor variables were screened using correlation analysis, scatter plots and seasonal variance tools of the SDSM model to determine the predictors that were strongly correlated with the predictands (daily minimum, maximum temperatures and precipitation in this case). Due to the decimal values in observed precipitation, it was transformed to fourth root function in order to produce a linear relationship. For better correlation among the predictors and predictands, the selected confidence level was 95 % with p-value of 0.05. The predictor variables with good correlation with the predictands were used for calibration and validation of the model with the monthly mode of simulation.

The performance of the SDSM during calibration and validation was assessed by comparing the Standard Deviation (SD) of the observed and simulated values along with Root Mean Square Error (RMSE) and Coefficient of Determination (R^2) which were calculated according to Eqs. 6.1, 6.2 and 6.3 respectively. The minimum deviation of the observed and simulated SD, lower value of RMSE and closer value of R^2 to 1 indicates the model is in good agreement with the observed variables. The screened predictor variables along with the A2 and B2 scenarios data from HadCM3 were used as input in the validated SDSM to generate the downscaled climate change scenarios for the province.

$$SD = \sqrt{\frac{1}{2}\left((x_i - \bar{x})^2\right)} \tag{6.1}$$

$$RMSE = \sqrt{\frac{1}{n}\sum_{i=1}^{n}(S_i - O_i)^2} \tag{6.2}$$

$$R^2 = \frac{\sum S_i \times Oi - \sum S_i \times \sum O_i}{\sqrt{\sum S_i^2 - (\sum S_i)^2} \times \sqrt{\sum O_i^2 - (\sum O_i)^2}} \tag{6.3}$$

where,

n number of observations
X_i observed/simulated value for a particular day of month
X mean value of observed/simulated variables for a month
S_i simulated climate variable for ith time
O_i observed climate variable for ith time

6.2.4 Crop Modeling

The AquaCrop v 4.0 was used to simulate the crop yield in this study. The AquaCrop is a crop water productivity simulation model developed by the Food and Agriculture Organization (FAO) of the United Nations; the model is the result and improvement of a key reference paper on agricultural yield responses to water (Doorenbos and Kassam 1979). The model estimates crop growth, given a set of climate and soil parameters, together with crop management. As the model was designed to assess crop response to water, it allows the evaluation of climate variability and change impact (changes in temperature, precipitation, relative humidity, wind speed, CO_2 concentrations, and reduced water availability) or environmental regulations (reduced water quotas) on crop yields. Various studies reported the ability of the model to simulate yield with good accuracy at different locations across the globe; making it convincing to use for the chosen study site.

AquaCrop calculates above ground biomass based on normalized water productivity and ratio of transpiration and reference evapotranspiration (Eq. 6.4). The yield is assumed as a function of reference harvest index and above ground biomass (Eq. 6.5).

$$B = k_{sb} \times WP^* \sum (T_r/ET_o) \tag{6.4}$$

$$Y = f_{Hi} \times HI_o \times B \tag{6.5}$$

where B is above ground biomass in t/ha, WP* is the normalized water productivity (gets adjusted for CO_2, synthesized yield production and soil fertility), (T_r/ET_o) is the ratio of crop transpiration and reference evapotranspiration, Y is yield in t/ha, f_{Hi} is the adjustment factor for heat, water and cold stress, HI_o is the reference harvest index. More details on AquaCrop can be found on Steduto et al. (2009) and Raes et al. (2009).

6.2.5 Agro-adaptation Measures

The following measures were evaluated as adaptation measures in order to overcome the projected impact of climate change: shifting transplanting dates and providing supplementary irrigation on rice yield under future climate. Simulations were carried out for CH207 cultivars at different transplanting dates ranging from 16 December to 24 February and 16 April to 25 June at an interval of 7 days for the winter and summer rice in the case of the three time windows. The current transplanting date is around 20 January and 21 May for winter and summer rice cultivation respectively. Additionally, simulations were also done for the future climate with supplementary irrigation of 4 applications of 20, 40, 60, 80 and 100 mm at an interval of 20 days starting from 20 days prior to flowering. In addition to the above mentioned adaptation measures which already exist in literature, we have further evaluated the influence of fertilizer amount and increase in splits of doses on rice yield under future climate. Simulations were carried out for rice yield with altered nitrogen, phosphorous and potassium inputs ranging from 0.5 to 2.5 times the recommended application rate of 100 kg/ha NH_4-N, 50 kg/ha P_2O_5 and 50 kg/ha K_2O for both summer and winter cropping. Moreover, simulations for two, three and four splits of fertilizer doses of the recommended amount were also carried out under future climate; to evaluate its influence. Literature also suggests that changing from traditional cultivars to heat tolerant varieties also serves as effective adaptation (Rezaei et al. 2013). Simulations were also carried out for two other heat resistant cultivars, TBR1 and OM6162, under future climate.

6.2.6 Methodological Limitations

The major assumptions of this study include that the area for rice will remain the same in the future. Also, other climatic parameters; specifically wind speed, and humidity were assumed to be the same in the future as in historical periods under the different scenarios. Furthermore, bias correction was not applied to the raw data of HadCM3 GCM which would have improved the screening of predictor variables and calibration of the SDSM. Despite the considerations mentioned above, it is emphasized that this study is extremely helpful in identifying tendencies and patterns although the figures may vary due to the limitations.

6.3 Results and Discussion

6.3.1 SDSM Calibration and Validation

The SDSM model was calibrated using temperature and precipitation data for the period of 1961–1990 whereas the validation was performed using data for 1991–2000 for the Tra My station. Daily maximum and minimum temperature and daily precipitation was downscaled for the station for impact assessment.

The first step for calibration was to screen the local scale predictor variables. Table 6.3 shows the predictors which have significant influence on the predictands. It is evident that for maximum and minimum temperature, mean temperature at 2 m has a very good correlation (>0.75). However, for precipitation, the predictor-predictand relationships are observed to be poor and therefore only three predictors with the highest partial correlation coefficient (r) are considered. It can also be observed that humidity has a significant correlation with precipitation (>0.25).

The calibration and validation of the model with the screened predictors suggests the model can simulate the observed variables in good agreement (Table 6.4).

Table 6.3 Summary of selected predictor variables and their respective predictands

Predictand	Predictors	Partial correlation coefficient (r)
Maximum temp	Mean temperature at 2 m	0.872
Minimum temp	Mean temperature at 2 m	0.796
Precipitation	Near surface relative humidity	0.253
	Mean temperature at 2 m	0.211
	Surface specific humidity	0.103

Table 6.4 Performance of the SDSM during calibration and validation

		Predictands		
		Maximum temperature	Minimum temperature	Precipitation
R^2	Cal	0.82	0.88	0.89
	Val	0.97	0.93	0.62
RMSE	Cal	0.51 °C	0.44 °C	47.19 mm
	Val	0.66 °C	0.84 °C	59.70 mm

Cal calibration, *Val* validation

Simulated temperature is in very good agreement with observed values; however the model failed to simulate monthly precipitation within an appreciable range. Further analysis of the monthly precipitation during validation of the model illustrates the shortfalls in simulated values existing throughout the year. Moreover, the highest deviation in observed and modeled values were for January. Nevertheless, based on R^2, RMSE and pattern of generated rainfall, it can be concluded that the model can simulate precipitation within an acceptable range. The obtained results can also be compared to those of Lines et al. (2006), Yang et al. (2012) and Charles et al. (2013), where simulation of precipitation is argumentative due to many regional driving factors but still fairly representative.

6.3.2 Projection of Future Climate

6.3.2.1 Estimation of Future Temperature

Simulation suggests that in the Quang Nam province, maximum temperature is expected to increase by 3.69 and 2.78 °C for A2 and B2 scenarios respectively by the 2080s relative to 1961–1990 (baseline period) (Table 6.5). A lower magnitude of increment 1.72 and 1.29 °C is projected for the 2080s for the corresponding scenarios, in case of minimum temperature (Table 6.5). Figure 6.3a, b represent the average monthly maximum temperature projected for the three time periods for A2 and B2 scenarios respectively. The analysis shows an insignificant variation in the trend of maximum temperature for both scenarios. However, an increase in temperature is observed from March to June for both scenarios and all time windows,

Table 6.5 Future projected changes in maximum and minimum temperature in the Quang Nam province

Time period	Maximum Temperature (T_{max}) (°C)				
	Baseline (1961–1990)	Scenario A2		Scenario B2	
		T_{max}	Change	T_{max}	Change
2020s	30.11	31.04	0.93	31.09	0.98
2050s		32.49	2.38	31.90	1.79
2080s		33.80	3.69	32.89	2.78
		Minimum Temperature (T_{min}) (°C)			
		Scenario A2		Scenario B2	
		T_{min}	Change	T_{min}	Change
2020s	21.60	21.95	0.35	21.99	0.39
2050s		22.70	1.10	22.41	0.81
2080s		23.32	1.72	22.89	1.29

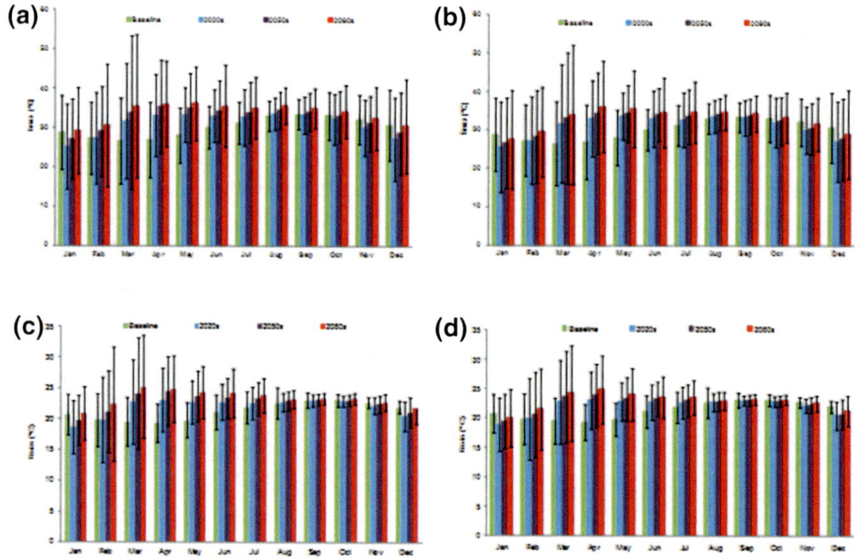

Fig. 6.3 Projected future average monthly maximum temperature under **a** A2 and **b** B2 scenarios and future average monthly minimum temperature under **c** A2 and **d** B2 scenarios (*Error bars* show variance of temperature across different years for various months)

indicating its implications in summer rice yield (discussed later). Furthermore, variation in the projected maximum temperature among inter time periods between the 2020s and 2080s are minor, with a maximum of 3.8 and 4.0 °C for March and April for A2 and B2 scenarios respectively.

A similar trend is also observed for average monthly minimum temperature where the projected maximum increase coincides with average monthly maximum

Table 6.6 Future projected changes in average annual precipitation in the Quang Nam province

Time period	Precipitation (Precp) (mm)				
	Baseline (1961–1990)	Scenario A2		Scenario B2	
		Precp	Change (%)	Precp	Change (%)
2020s	2655.44	2673.05	0.66	2703.92	1.83
2050s		2801.63	5.51	2747.65	3.47
2080s		2914.41	9.75	2804.70	5.62

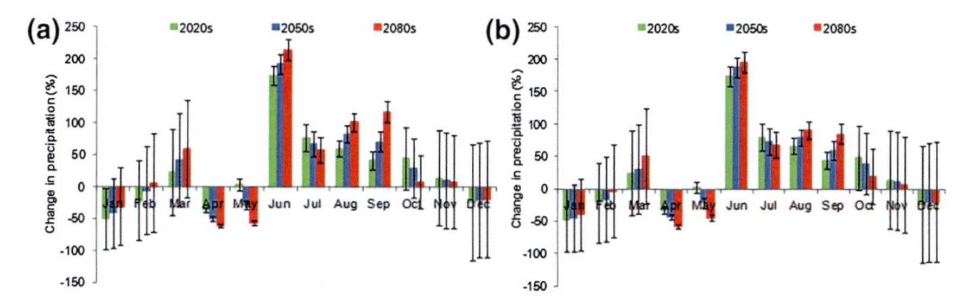

Fig. 6.4 Percentage change in projected monthly precipitation for **a** A2 and **b** B2 scenarios relative to baseline precipitation (*Error bars* show variance of precipitation change across different years for various months)

temperature. In addition, a lower elevation in minimum temperature is observed for all three periods relative to maximum temperature. However, it is interesting to note that for both scenarios, no change in average monthly minimum temperature is observed during August to December (Fig. 6.3c, d).

6.3.2.2 Projection of Future Precipitation

The projection of the future average annual precipitation under A2 and B2 scenarios suggests an increasing trend in precipitation. The maximum increase in precipitation for A2 and B2 scenarios is +9.75 and +5.62 % respectively for the 2080s relative to the baseline period (Table 6.6). Temporal analysis of the projected precipitation suggests an unambiguous variation with negative change in January, April, May and December for all scenarios and time periods (Fig. 6.4a, b). This will significantly influence the crop water availability in the future time periods (discussed later). A study on inter scenarios suggests an insignificant variation in the trend for all time periods except that higher magnitude is expected for the A2 scenario.

6.3.3 Projection of Climate Change Impact on Rice Yield

6.3.3.1 Sensitivity Analysis of the AquaCrop Model

Prior to conducting the calibration and validation of the crop model, sensitivity analysis of 12 parameters of the AquaCrop model was performed to select the most sensitive parameters for calibration. The input values of the parameters were adjusted by ±25 % relative to the default values and the simulations that were carried out. The response of the altered parameters in the outputs were noted and based on the criteria defined by Greets et al. (2009) they were grouped into high, moderate and low sensitive. It is observed that the canopy growth coefficient, water stress coefficient, stomatal stress coefficient and coefficient of HI at flowering and before flowering has a significant influence on yield (Table 6.7). In addition, yield is also observed to be moderately sensitive to the parameters, namely the canopy decline coefficient and water stress coefficient (upper).

6.3.3.2 Calibration of the AquaCrop Model

The AquaCrop model was calibrated for summer and winter cropping seasons by fine tuning the selected parameters from the sensitivity analysis to simulate yield and biomass, further comparing them to the observed ones from the research centres. Ten years (2001–2010) of available data was used for calibration of the model. Table 6.8 represents the calibrated values of the parameters for both cropping seasons whereas Table 6.9 presents the performance statistics of the model during the calibration procedure based on simulation of yield and biomass. It can be

Table 6.7 Sensitivity analysis of the parameters in the AquaCrop model

Input parameters	Sc (+25 %)	Sc (−25 %)	Sensitivity level
Maximum canopy cover	0.28	1.54	Low
Canopy growth coefficient	15.61	19.56	High
Canopy decline coefficient	12.17	9.65	Moderate
Maximum effective rooting depth	0.65	0.11	Low
Water stress coefficient (P_{upper})	6.54	8.69	Moderate
Water stress coefficient (P_{lower})	17.59	21.66	High
Water stress coefficient curve shape	0.39	2.14	Low
Stomatal stress coefficient (P_{upper})	15.65	18.29	High
Aeration stress coefficient	0.12	0.36	Low
Coefficient, HI increased by inhibition of leaf growth at flowering	16.39	17.25	High
Coefficient, HI increased due to inhibition of leaf growth before flowering	18.95	21.32	High
Canopy senescence stress coefficient (P_{upper})	0.25	0.22	Low

Table 6.8 Calibrated parameters of AquaCrop for CH207 cultivars of rice

Parameters	Summer	Winter	Unit
Canopy growth coefficient (CGC)	23.25	21.75	%/day
Canopy decline coefficient (CDC)	10.30	9.35	%/day
Water stress coefficient (WP_{upper})	0.75	0.62	% of TAW
Water stress coefficient (WP_{lower})	0.52	0.38	% of TAW
Stomatal stress coefficient (SP_{upper})	0.50	0.50	Unit less
Coefficient, HI increased by inhibition of leaf growth at flowering (HI_f)	0.82	0.75	Unit less
Coefficient, HI increased due to inhibition of leaf growth before flowering (HI_{bf})	0.23	0.28	Unit less

Table 6.9 Performance of the AquaCrop model during calibration

Cropping Season	Outputs	R^2	RMSE (t/ha)
Summer	Yield	0.84	0.32
	Biomass	0.79	0.69
Winter	Yield	0.92	0.26
	Biomass	0.84	0.58

noticed that the model simulates the yield and biomass in good agreement with those observed with appropriate parameterization.

Literature suggests that the AquaCrop model has been successfully extensively used in recent times for numerous crops due to its simplicity and robustness. Several researchers have applied the model for simulation of rice at various agro-ecological zones on a global scale. Shrestha et al. (2014) applied AquaCrop to simulate future rice yield and irrigation water requirements under climate change scenarios in Myanmar. In a separate study, Mainuddin et al. (2013) assessed the impact of climate change on rainfed rice in the lower Mekong Basin using AquaCrop. Although the study had a very coarse spatial resolution, the outcomes were satisfactory. Various other studies suggest that AquaCrop can be used as an effective tool in diverse climate conditions (Shrestha et al. 2013; Khoshravesh et al. 2013; Zhang et al. 2013).

6.3.3.3 Impact of Climate Change on Rice Yield

The average yield of rice for the periods 2020s, 2050s and 2080s were estimated for A2 and B2 scenarios. The yield was projected for summer (irrigated) and winter (rainfed) rice separately. Simulation results for summer cropping with full irrigation suggests increased yield for the 2020s and 2050s for both scenarios; however a decline in average yield is observed for the 2080s (Table 6.10). The highest projected maximum and minimum temperature during the flowering stage (July–

Table 6.10 Simulated rice yield and biomass for future climate in the summer and winter seasons

Period		CO_2 concentration	Summer			Winter		
			Yield (t/ha)	Change (%)	Biomass (t/ha)	Yield (t/ha)	Change (%)	Biomass (t/ha)
Baseline	2001–2010		5.42		15.38	5.21		14.89
A2	2020s	423	5.69	5.00	16.26	4.90	−5.97	14.00
	2050s	654	5.53	2.07	15.80	4.72	−9.42	13.49
	2080s	821	5.08	−6.26	14.51	4.01	−23.05	11.46
B2	2020s	411	5.78	6.66	16.51	5.14	−1.29	14.70
	2050s	496	5.57	2.79	15.91	4.83	−7.31	13.80
	2080s	529	5.32	−1.83	15.20	4.64	−10.96	13.26

August) leading to sterility of spikelet, reduced grain-filling duration and enhanced respiratory losses, can be attributed to the declining yield. In addition, results on inter scenario projections suggest both follow the same trend in yield projection, yet the B2 scenario is observed to have a better yield relative to A2. The projected increase in yield for the 2020s and 2050s can be attributed to the elevated CO_2 concentration in the atmosphere and its interaction with rice over temperature, although most crop models are unable to trace the interactions between yield and CO_2 concentration (Wassmann and Dobermann 2007).

Simulation winter (rainfed) rice yield shows a significant yield reduction under future time periods. The yield is expected to reduce by 5.97, 9.42 and 23.05 % for the 2020s, 2050s and 2080s compared to baseline yield under the A2 scenario (Table 6.10). A relative lower magnitude of 1.29, 7.31 and 10.96 % in yield reduction is observed for the B2 scenario for the corresponding periods. Projected reduction in precipitation during the transplanting and flowering stages relative to the baseline period could be the contributing factor to lower yield.

Furthermore, increase in temperature during the winter season led to raised evaporative demand of crops throughout the growing period, and an additional factor could be a reduction in precipitation, leading to unavailability of water. In summary, climate change will have a significant influence on the summer and winter rice yield if the current agricultural practice remains the same in future.

The outputs of this research can be compared to that of Kim et al. (2013) where their study suggests projection of yield varies both on a spatial and temporal scale. Korea expects an increase in rice yield of up to 22 % by the latter part of the century whereas depending on cultivar, it is also estimated to reduce. A positive indication is also observed for Harbin (China), due to climate change in the latter part of the century. A separate study by Kawasaki and Herath (2011) illustrates that projected rice yield in Northeast Thailand is specific to cultivars. Certain cultivars will benefit due to climate change; however, some will be negatively influenced. Further study on the Northeast of Thailand indicates that potential rice yield is expected to be higher for wetter periods (APN 2010). However, a comprehensive study of climate change impact on rainfed rice at a coarse spatial resolution in the lower Mekong river basin suggests that Vietnam and Cambodia are expected to have a significant reduction in yield in the 2050s relative to historical yield. The variability of projected rainfall in the region during the growing season has been attributed to the yield reduction (Mainuddin et al. 2013). In general, the region is expected to have a reduction in rice productivity due to the expected alteration of rainfall pattern and increase in temperature.

6.3.4 Evaluation of Agro-adaptation Measures

Several studies have validated that without adaptation and mitigation strategies, climate change will have a detrimental effect on agricultural production and economics (Smit and Skinner 2002). In order to identify the most suitable adaptation measures, it is of utmost importance to consider the climatic variables and evaluate the role of non-climatic factors that delicately influence agriculture. Altering the crop management practices, including changes to transplanting dates and application of supplementary irrigation has proven to be an efficient tool to counteract the negative impact of climate change (Babel et al. 2011; Jalota et al. 2012; Moradi et al. 2013). On the other hand, literature suggests that there is very little knowledge on the importance of fertilizer application rates and adjusting the number of split doses can also be applied as an adaptation approach under climate change. Furthermore, although changing the farming practice from traditional to heat and stress tolerant cultivars is also well documented, it has not been applied successfully in many regions of the world (Rezaei et al. 2013). Therefore the results of our study provide comprehensive information on possible agro-adaptation measures and the outputs can be used as guidance in other regions of the world to minimize the negative impact of climate change.

6.3.4.1 Altering the Transplanting Date

A simulation study on changing the transplanting date shows significant variations in the rice yield for both cropping seasons. From the simulations for the A2 scenario, it can be observed that early transplanting dates can reduce the yield for both seasons; however, late transplanting is observed to be beneficial in both cases. The maximum yield obtained when the rice is transplanted on 24 February for the winter season can increase the yield up to 20 % by the 2080s compared to the yield obtained by the current transplanting date of 20 January (Fig. 6.5a). Although results suggest a delay in the planting date approaches to the period of higher temperature, the increased evaporative demand of crop is eventually met by the observed increase in precipitation during the reproductive phase of the crop growth. A similar cause can also be attributed in the case of summer cropping where the maximum yield is obtained when the rice is transplanted on 11 June for the 2020s and 2050s and 18 June for the 2080s (Fig. 6.5b). Apparently, with a late transplanting date, a higher increase in yield is observed for summer cropping relative to winter, which can be attributed to higher available water in the period which is necessary during the flowering stage.

A similar result is obtained for simulation in the case of a B2 scenario and winter season which suggests a maximum yield (22.94 % for the 2080s) can be obtained by transplanting on 24 February relative to the yield by transplanting on the current date (20 January). Similarly, for the summer cropping season, transplanting on 11 June is optimum for the 2020s and 2080s with an increase of 26.72 and 22.86 % relative to the yield obtained by the current date of transplanting of 21 May. However, for the 2050s, a maximum yield is obtained by transplanting on 4 June (Fig. 6.5c, d). The lower magnitude of increase in minimum temperature, in addition to the higher precipitation expected during the growth phase with late transplanting is the contributing factor for higher yield in the winter season.

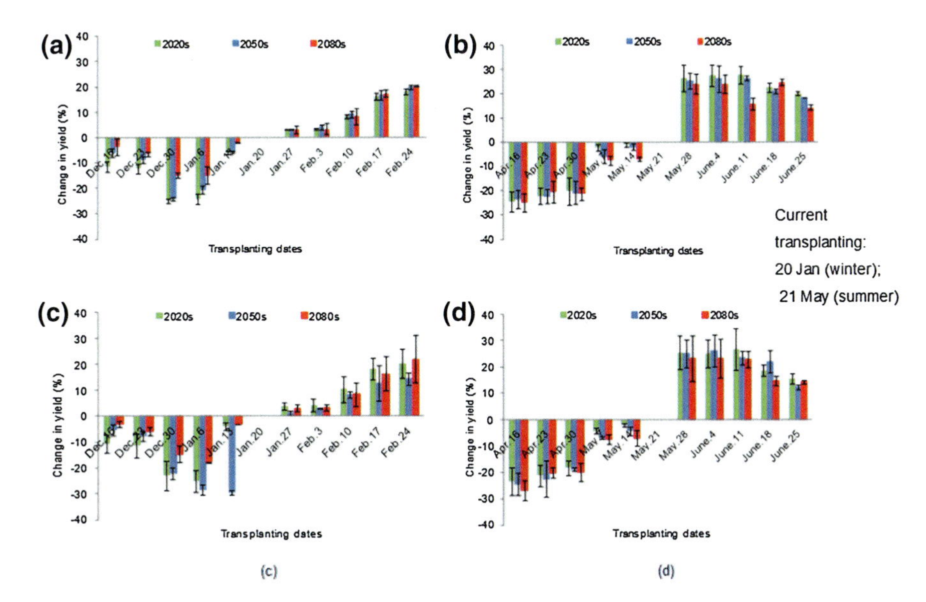

Fig. 6.5 Change in rice yield of CH207 cultivar with different transplanting dates **a** Winter and **b** Summer cropping season under the A2 scenario and **c** Winter and **d** Summer cropping season under the B2 scenario in the Quang Nam province (*Error bars* show variance of change in rice yield across different years for various transplanting dates)

However, escape from the critical stages (flowering and grain filling) of the higher temperature stress due to late transplanting can be the reason for an increase in yield for the summer season.

6.3.4.2 Introducing Supplementary Irrigation

Simulations were carried out assuming the application of supplementary irrigation by furrow method with an incremental rate of 20, 40, 60, 80 and 100 mm. Each irrigation level was applied four times at 20 day intervals starting from 20 days prior to the flowering stage in order to coincide with the critical stages of rice growth, flowering and grain filling. For the A2 scenario, the simulation shows a significant increase in yield for winter (rainfed) rice with the application of irrigation. It is observed that a total application of 400 mm can lead to a maximum yield of 24.13, 27.45 and 42.10 % for the 2020s, 2050s and 2080s compared to the rainfed yield for the corresponding time periods (Fig. 6.6a). As the summer rice is irrigated, providing additional water does not show any significant change of yield and hence it is not advised (Fig. 6.6b). Application of supplementary irrigation in the B2 scenario also suggests a higher yield can be achieved with irrigation in the winter season. With an application of 400 mm throughout the growing period of rice, potential yield can be enhanced by 30.45 and 32.81 % for the 2050s and 2080s

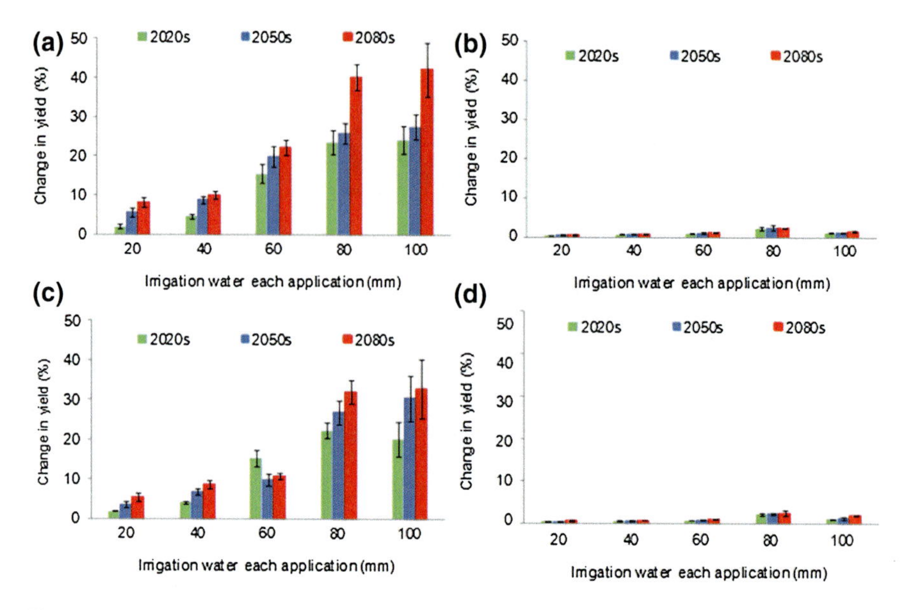

Fig. 6.6 Effect of supplemental irrigation on rice yield **a** Winter and **b** Summer cropping season for the A2 scenario and **c** Winter and **d** Summer cropping season for the B2 scenario (*Error bars* show variance of change in rice yield across different years for various irrigation water depth)

respectively. However, a relative lower input of 320 mm is noted to maximize the increase in yield by 23.05 %. In a case where the summer cropping season is analogous to the A2 scenario, a negligible increase is observed and therefore it is not suggested to provide additional irrigation (Fig. 6.6c, d). The ability of irrigation to adjust the canopy temperature irrespective of outside air temperature can be attributed to the increased yield against the potential heat stress during the winter cropping for both scenarios. In addition, with the current transplanting date for future climate, the evaporative water demand is unmet as a reduction in rainfall is observed (discussed earlier) and hence additional water application can enhance the yield up to its potential.

6.3.4.3 Changing Fertilizer Application Rate

In low fertile soil, proper nutrient management is of utmost importance. Our results show that a different application rate of NH_4-N, P_2O_5 and K_2O leads to different rice yields under future climate. An increase in yield is observed for a higher application rate of fertilizer although the magnitude varies as different time windows are considered. For instance, in the case of the 2020s, the highest yield is obtained for 1.5 times the application of the current application rate for both scenarios and cropping seasons. Similarly, for the 2050s and 2080s, 2 and 2.5 times the application rate is observed to give maximum yield (Fig. 6.7). The higher fertilizer

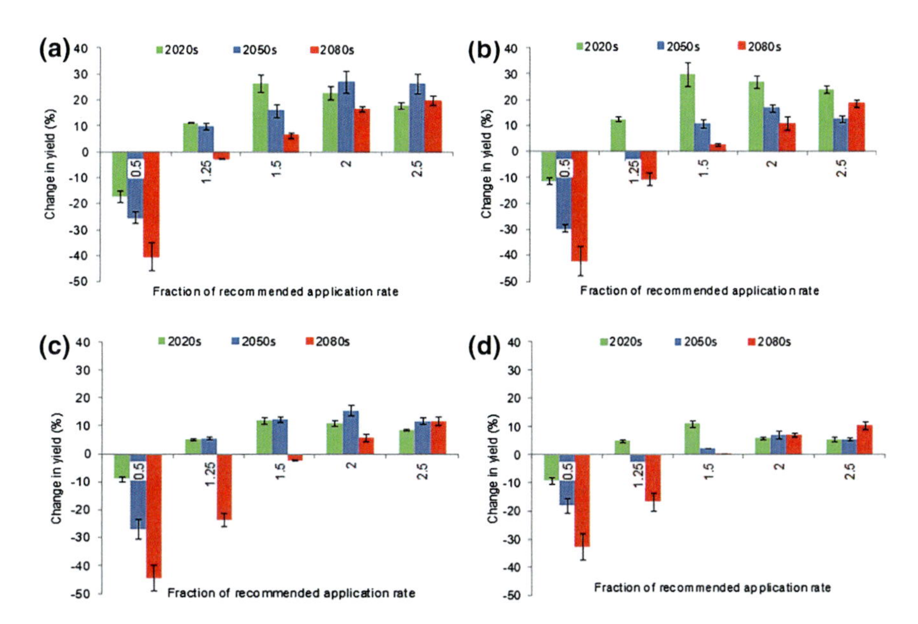

Fig. 6.7 Effect of different application rates of nitrogen-phosphorous-potassium (NPK) fertilizers on rice yield for **a** Winter and **b** Summer cropping under the A2 scenario and **c** Winter and **d** Summer cropping under the B2 scenario (*Error bars* show variance of change in rice yield across different years for various fractions of the recommended application rate)

input can be attributed to the positive charge of the nitrogen, phosphorous and potassium ions after disaggregation is easily absorbed by the soil colloids affecting the fertilizer efficiency use in the plants. In addition, expected higher rainfall in the future during the cropping period contributing to higher runoff and leaching fertilizers can also be claimed to be an attributing factor for higher fertilizer input. These results can also be compared to those of Babel et al. (2011), where a higher input of nitrogen fertilizers can maximize the potential rice yield in Northeast Thailand under climate change.

6.3.4.4 Changing Number of Fertilizer Doses

Farmers in developing nations generally follow the traditional practice of applying recommended fertilizer amounts to fields in a single dose at basal stage. However, we tried to simulate the yield in multiple split doses of equal amounts in 2 (10 days prior to both transplant and anthesis), 3 (10 days prior to transplant, panicle initiation and anthesis) and 4 (10 days prior to transplant, panicle initiation, anthesis and ripening) with the recommended amount of fertilizers for future climate for both seasons. It is evident that a higher split dose (4) application enhanced yield for the A2 scenario in the case of both cropping seasons relative to the current practice of a single dose prior to transplanting (Fig. 6.8a, b). However, in the case of the B2

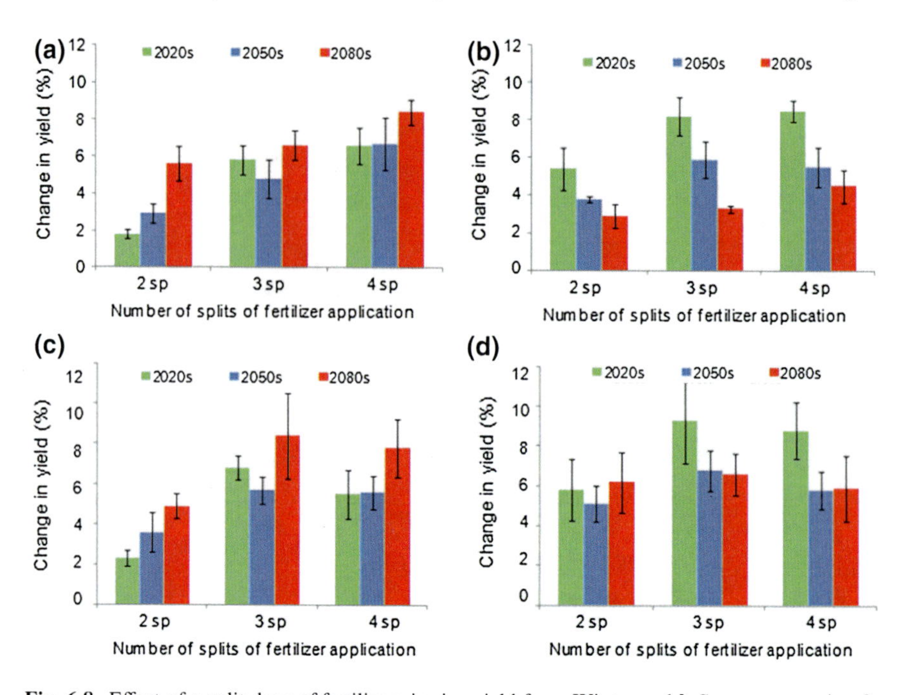

Fig. 6.8 Effect of a split dose of fertilizers in rice yield for **a** Winter and **b** Summer cropping for the A2 scenario and **c** Winter and **d** Summer cropping for the B2 scenario (*Error bars* show variance of change in rice yield across different years for the number of splits of fertilizer application)

scenario, the maximum yield is observed in the case of 3 split doses, although the variation corresponding to 4 split doses is insignificant for all the time windows (Fig. 6.8c, d). Providing fertilizers in smaller amounts accelerates the absorption rate of the Nitrogen-Phosphorus-Potassium (NPK) ions by the roots from the basal region. Moreover, providing doses of fertilizers prior to the critical stages of rice growth increases the photosynthesis rate and therefore enhances the growth of panicles and intensifies the number of grains per plant which ultimately leads to a higher yield.

6.3.4.5 Change in Cultivars

Developing rice cultivars for continued consistent spikelet development at higher temperature along with tolerance to submergence and water stress will have a positive influence on climate risk in rice yield. In this context, we simulated grain yield for two cultivars TBR1 and OM6162 under historical and future climate. The crop model was calibrated corresponding to the cultivars for the study site and parameters are presented in Table 6.11.

Table 6.11 Calibrated parameters of AquaCrop for TBR1 and OM6162 cultivars of rice

Parameters	TBR1		OM6162		Units
	Summer	Winter	Summer	Winter	
CGC	18.92	16.22	17.86	12.19	%/day
CDC	8.16	8.02	7.79	7.06	%/day
WP_{upper}	0.66	0.45	0.69	0.37	% of TAW
WP_{lower}	0.55	0.37	0.54	0.40	% of TAW
SP_{upper}	0.51	0.49	0.50	0.46	Unit less
HI_f	0.82	0.68	0.88	0.58	Unit less
HI_{bf}	0.26	0.22	0.21	0.26	Unit less

The simulated yield under future climate with current management practice for the two cultivars suggests a higher potential yield ranging from 9.58 to 20.73 % can be achieved for the winter season and 20 to 40.39 % for summer relative to the yield for CH207 for all time windows and A2 scenario (Fig. 6.9a, b). Similarly, a higher yield is also observed from 6.82 to 19.03 % for winter and 15.17 to 23.92 % for summer cropping in the case of the B2 scenario (Fig. 6.9c, d). It is also evident that for the baseline period, a larger yield can also be achieved with the heat tolerant varieties for both cropping seasons. This could be due to the projected decrease in temperature intensity in the future for the B2 scenario. Moreover, the heat resistant

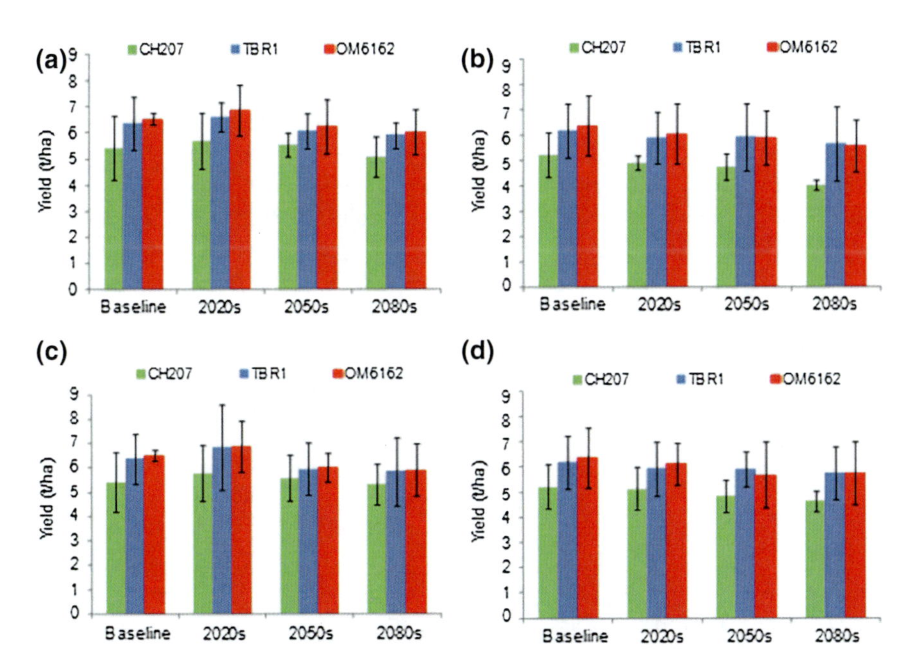

Fig. 6.9 Yield of different rice cultivars for future periods under the A2 scenario for **a** Winter and **b** Summer cropping and under the B2 scenario for **c** Winter and **d** Summer cropping (*Error bars* show variance of rice yield across different years for different cultivars)

varieties are less affected by spikelet injury, although there is an increase in temperature. The ability of these cultivars to change leaf orientation, transpirational cooling, alteration in the membrane lipid composition and short-term stress avoidance helps the plants to cope with the heat and water stress.

6.4 Conclusions

This study examines the potential impact of climate change on rice yield growth in winter and summer seasons along with an evaluation of agro-adaptation measures in order to overcome the projected impact of climate change in the Quang Nam province of Vietnam. The climate projection was done by downscaling the GCM data at province level using the Statistical Downscaling Model (SDSM) which indicates the future maximum and minimum temperature together with precipitation are expected to increase. Simulation of rice yield with future climate data suggests that rice cultivated in the winter season will suffer a significant yield loss ranging from 5.97 to 23.05 % and 1.29 to 20.96 % for A2 and B2 scenarios respectively. On the other hand, climate change is observed to be beneficial for rice cultivated in the summer season with an increase in yield (2.07 to 6.66 %) for the 2020s and 2050s, although a moderate reduction in yield (1.83 % to 6.26 %) is expected for the 2080s.

The study reveals that late shifting of transplanting dates to 18 June and 24 February relative to current planting dates of 21 May and 20 January for summer and winter rice respectively, enhances the potential yield under climate change. Also, 4 applications of 100 mm irrigation to winter rice increases yield significantly, relative to rainfed conditions. In addition, results suggest that applications of 2–2.5 times the current fertilizer amounts will increase the rice yield from 10.9 to 29.8 % in the future. Furthermore, 4 split doses of fertilizers enhances yield by 1.8 to 5.1 % of the yield relative to a single dose. Finally, shifting from the traditional cultivar CH207 to heat resistant cultivars TBR1 and OM6162 can enhance rice yield up to +9.58 to +40.39 % and +6.82 to +23.92 % under A2 and B2 scenarios respectively for future time periods. The results of this study will provide policy makers and key stakeholders with a decision support tool to optimize rice production under changing climate regimes in the Quang Nam province of Vietnam. Moreover, the outcomes can also be used as a guideline for adaptation measures under climate change scenarios in other parts of the world.

References

Adejuwon JO (2006) Food crop production in Nigeria II. Potential effects of climate change. Clim Res 32:229–245

APN (2010) Asia-Pacific Network for Global Change Research 2010 Climate change in Southeast Asia and assessment on Impact, Vulnerability and Adaptation on Rice Production and Water Resource. Project Reference Number: CRP2008-03CMY-Jintrawet

Babel MS, Turyatunga E (2014) Evaluation of climate change impacts and adaptation measures for maize cultivation in the western Uganda agro-ecological zone. Theor Appl Climatol. doi:10.1007/s00704-014-1097-z

Babel MS, Agarwal A, Swain DK, Herath S (2011) Evaluation of climate change impacts and adaptation measures for rice cultivation in Northeast Thailand. Clim Res 46:137–146

Bindi M, Olesen J (2000) Agriculture. In: Parry ML (ed) Assessment of potential effect and adaptations for climate change in Europe: the Europe ACACIA project. Jackson Environment Institute, University of East Angola, Norwich, p 324

Boote KJ (2011) Improving soybean cultivars for adaptation to climate change and climate variability. In: Yadav SS, Redden RJ, Hatfield JL, Lotze-Campen H, Hall AE (eds) Crop adaptation to climate change. Wiley-Blackwell, Oxford, (ch26)

Bryan E, Deressa TT, Gbetibouo GA, Ringler C (2009) Adaptation to climate change in Ethiopia and South Africa: options and constraints. Environ Sci Policy 12:413–426

Charles A, Bertrand T, Elodie F, Harry H (2013) Analog downscaling of seasonal rainfall forecasts in the Murray Darling basin. Monthly Weather Review, 141, 1099–1117 doi:http://dx.doi.org/10.1175/MWR-D-12-00098.1

Chhetri NB, Easterling WE (2010) Adapting to climate change: retrospective analysis of climate technology interaction in the rice-based farming system in Nepal. Ann Assoc Am Geogr 100 (5):1156–1176

CICS (2012) SDSM files, Canadian institute for climate studies, http://www.cccsn.ec.gc.ca/?page=sdsm. Accessed 3 Sep 2013

Dasgupta S, Laplante B, meisner C, Wheeler D, Yan J (2007) The impact of sea level rise on developing countries: a comparative analysis. world bank policy research working paper 4136, Washington DC

Dharmarathna WRSS, Herath S, Weerakoon SB (2014) Changing the planting date as a climate change adaptation strategy for rice production in Kurunegala district. Sri Lanka Sustain Sci 9 (1):103–111

Doorenbos J, Kassam AH (1979) Yield response to water. FAO irrigation and drainage paper no. 33, FAO, Rome, Italy, p. 193

FAO (2010) FAOSTAT food and agriculture organization. http://faostat.fao.org/site/291/default.aspx. Accessed 7 Mar 2012

Giorgi F, Mearns LO (1991) Approaches to the simulation of regional climate change: a review. Rev Geophys 29:191–216

Gouache D, Bris XL, Bogarda M, Deudone O, Pagéd C, Gatee P (2012) Evaluating agronomic adaptation options to increasing heat stress under climate change during wheat grain filling in France. Euro J Agron 39:62–70

Greets S, Raes D, Garcia M, Miranda R et al (2009) Simulating yield response of quinoa to water availability with AquaCrop. Agron J 101:498–508

Iizumi T, Yokozawa M, Nishimori M (2011) Probabilistic evaluation of climate change impacts on paddy rice productivity in Japan. Climatic Change 107:391–415

IPCC (2000) IPCC Working Group III, 2000. IPCC Special Report Emissions Scenarios. Intergovernmental Panel on Climate Change, Cambridge University Press, Cambridge

IPCC (2007) Climate change 2007: impacts, adaptation and vulnerability. Contribution of Working Group II to the 4th assessment report of the intergovernmental panel on climate change. Cambridge University Press, Cambridge

Jalota SK, Kaur H, Ray SS, Tripathi R, Vashisht BB, Bal SK (2012) Mitigating future climate change effects by shifting planting dates of crops in rice-wheat cropping system. Reg Environ Change 12:913–922

Kawasaki J, Herath S (2011) Impact assessment of climate change on rice production in Khon Kaen province, Thailand. J Int Soc SE Asian Agric Sci 2:14–28

Khoshravesh M, Mostafazadeh-Fard B, Heidarpour M, Kiani AR (2013) AquaCrop model simulation under different irrigation water and nitrogen strategies. Water Sci Tech 67(1):232–238

Kim HY, Ko J, Kang S, Tenhunen J (2013) Impacts of climate change on paddy rice yield in a temperate climate. Glob Change Biol 19:548–562
Lines GS, Pancura M, Lander C (2006) Building climate change scenarios of temperature and precipitation in Atlantic Canada using SDSM. Science Report Series 2005-9, Meteorological Service of Canada, Atlantic Region, Dartmouth N.S. B2Y 2N6, Canada
Luo Y, Jiang Y, Peng S, Cui Y, Khan S, Li Y, Wang W (2013) Hindcasting the effects of climate change on rice yields, irrigation requirements, and water productivity. Paddy Water Environ. doi:10.1007/s10333-013-0409-8
Mainuddin M, Kirby M, Hoanh CT (2011) Adaptation to climate change for food security in the lower Mekong Basin. Food Sec 3(4):433–450
Mainuddin M, Kirby M, Hoanh CT (2013) Impact of climate change on rainfed rice and options for adaptation in the lower Mekong Basin. Nat Hazards 66(2):905–938
Meza FJ, Silva D, Vigil H (2008) Climate change impacts on irrigated maize in Mediterranean climates: evaluation of double cropping as an emerging adaptation alternative. Agric Syst 98:21–30
Mishra A, Siderius C, Aberson K, van der Ploeg M, Froebrich J (2013) Short-term rainfall forecasts as a soft adaptation to climate change in irrigation management in North-East India. Agric Water Manag 127:97–106
MONRE (2009) Climate change, sea level rise scenarios for Vietnam. Ministry of Natural Resources and Environment, Hanoi, Vietnam
Moradi R, Koocheki A, Mahallati MN, Mansoori H (2013) Adaptation strategies for maize cultivation under climate change in Iran: irrigation and planting date management. Mitig Adapt Strateg Glob Change 18:265–284
Nguyen T (2006) Poverty, poverty reduction and poverty dynamics in Vietnam. Background paper for the Chronic Poverty Report 2008–09. Chronic Poverty Research Centre, Manchester
Nguyen HN, Vu KT, Nguyen XN (2008) Flooding in Mekong river delta, Vietnam. Human Development Report 2007/2008. United Nations Development Program, Hanoi
Poudel S, Kotani K (2013) Climatic impacts on crop yild and its variability in Nepal: do they vary across seasons and altitudes? Climatic Change 116(2):327–355
Raes D, Steduto P, Hsiao TC, Fereres E (2009) AquaCrop: the FAO crop model to simulate yield response to water: II. Main algorithms and software description. Agron J 101:438–447
Ray DK, Ramankutty N, Mueller ND, West PC, Foley JA (2012) Recent patterns of crop yield growth and stagnation. Nat Commun. doi:10.1038/ncomms2296
Rezaei EE, Gaiser T, Siebert S, Ewert F (2013) Adaptation of crop production to climate change by crop substitution. Mitig Adapt Strateg Glob Change. doi:10.1007/s11027-013-9528-1
Rosenzweig C, Tubiello FN (2007) Adaptation and mitigation strategies in agriculture: an analysis of potential synergies. Mitig Adapt Stateg Glob Change 12:855–883
Ruosteenoja K, Carter TR, Jylhä K, Tuomenvirta H (2003) Future climate in world regions: an intercomparison of model-based projections for the new IPCC emissions scenarios. The Finnish Environment Institute report, Helsinki, Finland
Shrestha S, Gyawali B, Bhattarai U (2013) Impacts of Climate Change on Irrigation Water Requirements for Rice-Wheat Cultivation in Bagmati River Basin. Nepal J Wat Clim Change 4 (4):422–439
Shrestha S, Thin NMM, Deb P (2014) Assessment of climate change impacts on irrigation water requirement and rice yield for Ngamoeyeik Irrigation Project in Myanmar. J Wat Clim Change, In Press
Smit B, Skinner MW (2002) Adaptation options in agriculture to climate change: a typology. Mitig Adapt Strateg Glob Change 7:85–114
Soora NK, Aggarwal PK, Saxena R, Rani S, Jain S, Chauhan N (2013) An assessment of regional vulnerability of rice to climate change in India. Clim Chan 118(3–4):683–699
Steduto P, Hsiao TC, Raes D, Fereres E (2009) AquaCrop—the FAO crop model to simulate yield response to water: 1 Concepts and underlying principles. Agron J 101:426–437
Stocker TF, Qin D, Plattner GK, Alexander LV, Allen SK, Bindoff NL, Bréon FM, Church JA, Cubasch U, Emori S, Forster P, Friedlingstein P, Gillett N, Gregory JM, Hartmann DL, Jansen

E, Kirtman B, Knutti R, Kumar KK, Lemke P, Marotzke J, Masson-Delmotte V, Meehl GA, Mokhov II, Piao S, Ramaswamy V, Randall D, Rhein M, Rojas M, Sabine C, Shindell D, Talley LD, Vaughan DG, Xie SP (2013) Technical Summary. In: Stocker TF, Qin D (eds) Climate change 2013: the physical science basis. Contribution of working group I to the fifth assessment report of the intergovernmental panel on climate change. Cambridge University Press, Cambridge

Tao F, Zhang Z (2010) Adaptation of maize production to climate change in North China plain: quantify the relative contributions of adaptation options. Eur J Agron 33:103–116

Tao F, Zhang Z, Zhang S, Zhu Z, Shi W (2012) Response of crop yields to climate trends since 1980 in China. Clim Res 54:233–247

Tingem M, Rivington M (2009) Adaptation for crop agriculture to climate change in Cameroon: turning on the heat. Mitig Adapt Strateg Glob Chang 14:153–168

Vu L, Glewwe P (2008) Impacts of rising food prices on poverty and welfare in Vietnam. Working Paper Series No. 13. Development and Policies Research Center (DEPOCEN), Hanoi, Vietnam

Wassmann R, Dobermann A (2007) Climate change adaptation through rice production ion regions with high poverty levels. J Semi-Arid Tropical Agric Res 4(1). Available from: http://www.icrisat.org/journal/SpecialProject/sp8.pdf Accessed 11 Feb 2014

Wassmann R, Jagadish SVK, Sumfleth K, Pathak H, Howell G, Ismail A, Serraj R, Redona E, Singh RK, Heuer S (2009) Regional vulnerability of climate change impacts on Asian rice production and scope for adaptation. In: Advances in Agronomy, Elsevier, pp 91–133

Wilby RL, Dawson CW (2007) SDSM 4.2—a decision support tool for the assessment of regional climate change impacts, user manual

Yang JC, Zhang JH (2006) Grain filling of cereals under soil drying. New Phytol 169:223–236

Yang T, Li H, Wang W, Xu CY, Yu Z (2012) Statistical downscaling of extreme daily precipitation, evapotranspiration and temperature and construction of future scenarios. Hydrol Processes 26:3510–3523

Zhai, F, Zhuang J (2009) Agricultural impact of climate change: A general equilibrium analysis with special reference to Southeast Asia. Asian Development Bank institute working paper series 131. Asian Development Bank, Manila, the Philippines

Zhang W, Liu W, Xue Q, Chen J, Han X (2013) Evaluation of the AquaCrop model for simulating yield response of winter wheat to water on the southern Loess Plateau of China. Water Sci Tech 68(4):821–828

Printed by Publishers' Graphics LLC
DBT141027.03.05.331